# EURO Advanced Tutorials on Operational Research

**Series Editors**

M. Grazia Speranza, Brescia, Italy
José Fernando Oliveira, Porto, Portugal

More information about this series at http://www.springer.com/series/13840

Carlos Henggeler Antunes • Maria João Alves •
João Clímaco

# Multiobjective Linear and Integer Programming

Springer

Carlos Henggeler Antunes
University of Coimbra
Coimbra, Portugal

Maria João Alves
University of Coimbra
Coimbra, Portugal

João Clímaco
INESC Coimbra
Coimbra, Portugal

ISSN 2364-687X ISSN 2364-6888 (electronic)
EURO Advanced Tutorials on Operational Research
ISBN 978-3-319-28744-7 ISBN 978-3-319-28746-1 (eBook)
DOI 10.1007/978-3-319-28746-1

Library of Congress Control Number: 2016932386



Printed on acid-free paper

This Springer imprint is published by Springer Nature
The registered company is Springer International Publishing AG Switzerland

# Preface

In classical Operational Research, the decision maker's preferences are modeled a priori; that is, all the values are aggregated in a single objective function (or criterion), aimed at determining the *optimal* solution to the problem. However, it is currently recognized that these approaches are too reductive, being inadequate to address many real-world problems. In these problems, multiple perspectives should be taken into account to evaluate the merits of potential solutions; i.e., the decision maker is generally interested not just in minimizing the cost but also in maximizing the system reliability, minimizing the environmental impacts, etc. Approaches that make an a priori aggregation of the multiple perspectives cannot duly capture the conflicting nature of the objective functions, which make operational evaluation aspects of distinct nature and impair the exploitation of trade-offs among them. Therefore, multiobjective optimization models, which include explicitly the multiple evaluation aspects as distinct objective functions, enable to adequately capture the essential characteristics of real-world problems and improve their perception by decision makers. The concept of nondominated solution is the key concept in multiobjective optimization: that is, a feasible solution for which no other feasible solution exists improving all objective function values simultaneously. In this setting, the multiobjective optimization problem is defined as the choice, among the nondominated solution set, of a solution, or a reduced set of solutions for further screening, which reveals to be an acceptable compromise outcome as a result of the decision support process taking into account the decision maker's preferences. These preferences should not be understood as a preexisting and stable entity in the operational framework of such process, but they are subject to evolve as new information is gathered about the characteristics of nondominated solutions in different regions of the search space, which is determined by the mathematical model and the incorporation of additional elements derived from the preferences expressed. The decision support process using multiobjective optimization models is therefore based on interactive cycles of computation of nondominated solutions, evaluation, and possible change of preferences in face of new information. This information results from new solutions computed and their confrontation with information previously gathered, having in mind the "convergence" (not based on any type of aggregation functions previously developed) to a final solution that establishes an acceptable compromise between the competing objective functions.

This book aims at providing an entrance door into linear and integer multiobjective optimization, and it is primarily intended for undergraduate and graduate students in engineering, management, economics, and applied mathematics. It starts (Chap. 1) by introducing the motivation and interest of explicitly considering multiple objective functions in optimization models. Problem formulation, definitions, and basic concepts are then presented in Chap. 2, followed by the exposition of techniques to compute nondominated solutions and the role of preference information in those scalarizing techniques (Chap. 3). Chapter 4 is devoted to interactive methods, in which some methods representative of different search strategies are presented. In Chap. 5, a guided tour of the *iMOLPe—interactive MOLP explorer* software is presented, which was developed by the authors to deal with multiobjective linear programming problems. Chapter 6 deals with multiobjective integer and mixed-integer linear programming. It includes a literature review concerning the most relevant approaches, followed by the presentation of an interactive reference point method developed by the authors.

This book focuses on multiobjective linear, integer, and mixed-integer programming. These topics are adequate to get into multiobjective optimization, as they enable to shed light on both the extension of the traditional single objective linear, integer, and mixed-integer programming models (broad topics usually studied in Operational Research courses) and the different paradigm that is at stake when multiple objective functions are explicitly considered.

This is a textbook rooted on our experience of more than 25 years of research and teaching of multiobjective optimization, and it results from our conviction of the increasing importance of this topic in teaching Operational Research as a scientific basis to support the process of making more informed and better decisions.

Coimbra, Portugal
October 2015

Carlos Henggeler Antunes
Maria João Alves
João Clímaco

# Contents

# Chapter 1
# Introduction

Operational Research (OR) has developed as a scientific discipline during the decade of 1950, having created since then the (false) expectative that it would end up developing adequate methods and techniques for solving at least most decision problems faced at different levels, in industry and service sectors. In the beginning of the decade of 1970, the growing complexity of the economical and social environment and the swift cadence of technological innovation, particularly in the information and communication domains, made clear that the progress depended even more on the adoption of innovative planning and management procedures, narrowing the gap between the technological component and the methodological component of the production system. In these circumstances, the traditional quantitative methods of OR only were not able to suit themselves to the resolution of many problems.

It was in this context that Geoffrion (1983) wrote an article entitled "Can Management Science/Operations Research evolve fast enough?". May be there is not yet a definitive answer to this question. However, the accomplishments in several fields of interdisciplinary nature allow us to face the future with optimism, with new increasingly stimulating intellectual challenges arising. OR, understood under the perspective of the science and art of decision support, makes an appeal for the conjugated use of modern techniques of information systems, sophisticated human-computer interfaces, quantitative methods and algorithms, new modeling techniques, artificial intelligence techniques and certain disciplines usually included in the so-called human and social sciences, namely cognitive psychology and sociology of organizations.

In this context, the study of models with multiple, incommensurate and conflicting axes of evaluation of the merits of potential courses of action is a topic of utmost importance. In fact, real-world problems are intrinsically of multiobjective nature, being single objective approaches reductive in most cases.

In this tutorial, we begin by making the bridge between single objective linear programming (LP) models, for which the computation of the optimum is a "mere"



C. Henggeler Antunes et al., *Multiobjective Linear and Integer Programming*,
EURO Advanced Tutorials on Operational Research,
DOI 10.1007/978-3-319-28746-1_1

technical issue, and models that explicitly consider multiple objectives, thus questioning the optimality paradigm, which assumes a relation of comparability between the alternatives and the transitivity of the comparisons. In the single objective case, models translate just partially the aggregation of the decision maker's (DM's) preferences, leaving for posterior analysis the most controversial options leading to decision making. Multiobjective linear programming (MOLP) appears as a natural extension of LP, allowing to make the counterpoint between the normative nature of the single objective model and the symbiosis between quantitative and qualitative aspects of the decision process.

When about three decades ago we were involved in a power generation expansion planning problem using an LP model, we were confronted with the advantages of considering explicitly multiple conflicting objective functions: the global cost, a function associated with the reliability of the supply system and a function penalizing environmental impacts. The study of this problem led us not only to use diverse MOLP approaches, but also to the development of a new interactive tool called TRIMAP.

Optimization approaches, which were often developed by the stimulus of applications, had its roots in economic theory. For example, the simplex method presented in 1947 by Dantzig for solving LP problems, emerged in the sequence of works developed by well-known economists, such as Leontief, who developed the input–output model, Koopmans, who proposed a transportation model, Von Neumann and Morgenstern, creators of game theory. The relative simplicity and the efficacy of the simplex method broadened out the perspectives of the application of LP to actual problems in industry, services and public administration. During the decade of 1950, this impulse was decisive for the development of the main foundations of mathematical programming. Theoretical and methodological advances of the mathematics of optimization, the development of computing capability and the success of some applications led the heralds of the neoclassical theory of organizations to overvalue the use of mathematical optimization models. For the defenders of this theory, all models are based on an objective function, a value function or a utility function, which should be optimized. The formulation of the objective function is considered a minor problem; for example, Hitch (1953) referred to the problem of the criterion (objective function) choice as the simplest one in OR. The monetary unit appeared as the only measure of social benefit.

The difficulty of the DMs in participating in the formulation of the objective function and the lack of aspects of social nature, not quantifiable in monetary units, justified that already in the decade of 1950 several authors proposed the utilization of other types of models. For Keen (1977), the complexity of the actual problems arising in modern developed societies is essentially marked by multiple objectives, and the DM is often confronted more with the need of arbitrating the conflict between the objectives than the search for optimal solutions. Thus, a new branch of mathematical programming emerged—multiobjective programming (MOP) devoted to models in which multiple objective functions are explicitly considered.

It is opportune to refer that, under the common designation of multiple criteria approaches, two distinct branches appear in the specialized literature:

(a) methods for decision support with multiple attributes;
(b) methods for decision support with multiple objectives.

The first designation generally refers to the selection, ranking or categorization methods dealing with a finite set of alternatives, which are explicitly known a priori. The second designation is concerned with problems in which the alternatives are implicitly defined by a set of constraints. This tutorial is devoted to these latter problems, and within these to the particular cases of linear, integer and mixed integer programming with multiple objective functions.

Similarly to the case of mathematical programming with a single objective function, the roots of mathematical programming with multiple objectives dive in the economic theory. Pareto, in 1906, defined a concept that revealed fundamental for mathematical programming with multiple objectives, the concept of Pareto optimal solution (also named efficient, nondominated or noninferior solution). A solution is nondominated whenever there is no other feasible solution that simultaneously improves all the objective function values, i.e., improving an objective entails deteriorating, at least, one of the other objective function values.

Von Neumann and Morgenstern (1947), in their classical book on game theory and economic behavior, referred to the need of using more than one objective function in the following terms: "*The optimization problem in the context of a market economy is not certainly a problem of optimization, but a disconcerting and peculiar mixture of several conflicting problems of optimization …*". Koopmans introduced the notion of efficient solution in the context of the analysis of linear models of production in the monograph "*Analysis of Production as Efficient Combination of Activities*" (1951), one of his main works for which he was bestowed, jointly with Kantorovich, the Nobel Prize in Economics in 1975. Kuhn and Tucker (1951) considered for the first time several objective functions in mathematical programming models. They stated the conditions for *noninferiority* in *vector optimization* problems.

Before the 1960s there was not a systematic research in this field, although some publications can be traced. Geoffrion published two important articles about this topic (Geoffrion 1967, 1968). Since then, thousands of papers on mathematical programming with multiple objectives were published, of theoretical, algorithmic, computational and application nature. More recently, not only several international conferences and streams with major OR events devoted to this area have been held and working groups were created, but also books were published offering overviews of the state of the art.

Let us return to our energy planning problem and the options that we had to make when, after developing a suitable multiobjective linear programming model, we began studying it with the aim of supporting the DM. It was then necessary to find the adequate methods of analysis to obtaining/learning the DM's preferences regarding the cost, environmental impact and reliability objective functions.

Traditionally, the MOP methods are classified into three categories, according to the process used for aggregating the DM's preferences, with the aim of selecting the best compromise solution:

- methods where an a priori aggregation of preferences is made;
- methods of progressive articulation of preferences (interactive methods);
- methods in which no articulation of preferences is made (generating methods).

In the methods of the first type, although several objectives are explicitly modeled, the aggregation of preferences is made before any computation stage and the problem is a priori transformed into a single objective problem, for example through the construction of an utility function. The special characteristics of the geometry of the feasible polyhedron associated with our case study made particularly inadequate the application of methods in which decisions were taken a priori, without the DM having the possibility of realizing the consequences of other choices. We thought then that the most adequate approach would be the use of methods in which there is no articulation of preferences, being this task carried out *a posteriori* by the DM after the nondominated solution set has been fully characterized.

These generating methods of all nondominated solutions correspond, in the case of MOLP, to the extension of the simplex method for determining all the nondominated vertices of the feasible polytope, having the possibility of determining also the nondominated edges and faces of higher dimension. The first generating algorithms dedicated to MOLP were developed in the beginning of the decade of 1970, proposed by Evans and Steuer (1973) and Yu and Zeleny (1975). In our study we opted for the Yu and Zeleny's method and we published a first article on this study in 1981 (Clímaco and Almeida 1981). However, later on, when we carried out experiments using a model with a higher number of variables and constraints to obtain more realistic results, the issues associated with generating methods started to emerge. Above certain dimensions, the computational effort became impracticable. Even when it was possible to compute all the nondominated vertices, we verified that this involved a very high computational effort that was not compensated by the quality of the information offered to the DM. In fact, the DM revealed incapable of making a substantiated choice when confronted with a vast quantity of information, that is, with a very high number of nondominated solutions, mainly because many of them just showed smooth variations in the objective function values.

In many real world problems, the geometry of the feasible region leads to sub-regions of nondominated solutions where the variations of the objective function values are very smooth and other sub-regions where steep variations occur. Also, the DM often prefers nondominated solutions located on nondominated faces rather than extreme points because, in general, those solutions display a more balanced compromise between the competing objectives, which further complicates the study from the computational point of view.

Our energy planning case study led us to conclude that only methods of progressive articulation of preferences, in which interaction with the DM occurs, would allow to overcome the difficulties associated with value/utility function based methods and generating methods in order to offer useful information to the DM. Interactive methods encompass two essential phases: computation of nondominated solution(s) and dialogue between the DM/analyst and the computational tool implementing the method. These phases are alternately repeated until a stopping condition is reached. In general, due to the complexity of the methods, the communication between the DM and the computer is mediated by an analyst with expertise on methods and computational tools. The dialogue phase is regulated through a communication protocol, and the information elicited from the DM (in face of the solutions obtained in the previous computation phase) is used to prepare the subsequent computation phase to obtain a new solution, expectedly more in accordance with those expressed preferences. In MOP each computation phase consists in the resolution of a surrogate single objective problem (in some methods more than one problem is solved), so that the optimization of a scalarizing function built using the information elicited from the DM leads to a nondominated solution. A final compromise solution as the outcome of the decision support process should belong to the nondominated solution set (considering the model is sufficiently accurate to represent the main features of the real world problem).

As it was emphasized by Vanderpooten and Vincke (1989), it is essential that, on one hand, the computational effort is not too high in the computation phase and, on the other hand, the questions asked to the DM are simple and understandable. The first requirement is indispensable to guarantee that the decision process is truly interactive regarding the time expected for obtaining a new solution. The second one impacts on the quality of the solutions that are successively proposed to the DM, avoiding wrong indications in the dialogue phase. Therefore, in an interactive decision support process a cycle of proposals and successive reactions proceeds until a satisfactory compromise solution is identified.

The exploitation of interactivity in decision support processes based on multiobjective models and methods should respond to the following key questions:

- Does it make sense to mention the search for the *optimal solution* of any utility function that is assumed to exist, but for which an analytical representation is not known (that is, the convergence of the interactive process to the optimum of the implicit utility function)?
- What are the stopping conditions of the algorithm?
- What is the meaning of the final solution chosen?

The most well-known interactive methods in MOP are based on the local information collected from the computation phase, in which one or several scalarizing functions are optimized. This information is used to make questions to the DM, which allows obtaining the necessary inputs to prepare a new computation phase, in general in terms of parameters, such as weights, levels of aspiration, minimum acceptable values for the objective functions. Vanderpooten (1989) called the information obtained in this way preferential information, and the parameters that allow obtaining it are called preference parameters. The compromise between the complexity and the richness of the desirable information, and the capability of the DM to answer the questions formulated, is a subtle problem that transcends technical questions of mathematical programming, implying the consideration of studies about rationality in decision processes.

At this point it is relevant distinguishing two markedly antagonist attitudes in the development of interactive methods (for details see Roy 1987). The first way, that Roy calls "*la voie du realisme: l'attitude descriptive*", interpreted in terms of MOP can be described as: there is a DM's utility function whose optimization would lead to the optimal compromise solution in face of the multiple objectives. However, this utility function is not explicitly known by the DM, being the interactive process responsible for its discovery using dialogue protocols of algorithmic nature. The best optimal solution selected does not depend, therefore, on the evolution of the interactive process, being a primitive of the DM for the problem under study. Hence, an interactive method allows discovering the DM's utility function and consequently the convergence to its optimum. As we shall see in Chap. 4, this approach assumes that the DM is coherent with the answers to the questions that are asked by the communication protocol of the method. In some methods these questions embody a dichotomy; that is, each answer leads to a mutually exclusive division of the search space, which implies that henceforth there are solutions that are never again considered.

The application of this type of approaches to our case study revealed problematic since the DM was successively confronted with difficult options, due to the geometry of the nondominated region, either by corresponding to "jumps into the dark", when there were sharp variations of the objective function values, or being difficult to make options in areas with very smooth variations. Moreover, these approaches do not foster learning during the decision process, i.e., enabling the DM to grasp the characteristics of the problem and unveil the trade-offs at stake, with consequences on the creation of his/her system of preferences. Our experience confirms the opinions of authors as French (1984) ("*I believe that good decision aid should help the decision maker explore not just the problem, but also himself. It should bring to his attention possible conflicts and inconsistencies in his preferences so that he can think about their resolution*") and Lewandowski and Wierzbicki (1988) ("*A rational decision does not have to be based on all the available information, nor does it have to be optimal. It should only take into account the possible consequences of the decision and be intended not to be detrimental to the values and interests of the decision maker.*"). Learning during the interactive analysis is stimulated through the trial and error process. For example, passing twice by the same solution during the interactive process would not be "allowed" under the point of view of the convergence to the optimum of an utility function, while in learning-oriented approaches it is a natural fact, being even possible that the DM does not react the same way when confronted again with an already previously proposed solution.

These studies encouraged the development of an interactive environment devoted to linear programming with three objective functions, named TRIMAP, offering the capabilities for a holistic view of the nondominated solution set as the basis for then proceeding to a more focused search (Clímaco and Antunes 1987, 1989). TRIMAP is based on a selective and progressive learning of the nondominated solution set. The goal is not to converge to the optimal solution of any utility function, but rather to support the DM in the elimination of the subset of nondominated solutions that reveal not having practical interest. There are no irrevocable decisions during the interactive process, allowing the DM to review previous options. The consideration of three objective functions enables to use graphical means that are particularly adequate for the dialogue with the DM/analyst, simplifying the dialogue and increasing his/her information processing capability. Using the Feyerabend's terminology (1975), *open communication* is privileged and this interactive environment reveals adequate not just for the analysis of the problem but also for evaluating the several solutions by contrast. This is an important feature to overcome common problems in interactive procedures, such as "anchoring" in the first proposed solutions.

The interactive decision support process should be understood as a learning and constructive process in which the DM gathers information about the nondominated solution set and progressively shapes his/her preferences so that a final compromise solution may emerge.

MOP gained an increasing acceptance to provide decision support in real world problems in which multiple, incommensurate and conflicting axes of evaluation of the merits of the potential solutions should be explicitly considered. For instance, Greco et al. (2005, 2016) offer reviews of applications in location, finance, energy planning, telecommunication network planning and design, and sustainable development.

The topics addressed in this book have been taught by the authors in courses of Operational Research for undergraduate and graduate students in engineering, management, economics and applied mathematics to introduce MOP models and methods. It is our conviction that this is the adequate path for enhancing the students' understanding of the main issues at stake in multiobjective optimization, before progressing to more technically demanding topics or meta-heuristic approaches required by nonlinear characteristics or combinatorial nature of some problems (Deb 2001; Coello et al. 2002).

In Chap. 2 the formulation and definitions in multiobjective linear programming, integer programming, mixed integer programming and nonlinear programming are presented. The definitions of efficient and nondominated solutions, weak and proper efficient/nondominated solutions, supported and unsupported efficient/nondominated solutions, are presented and illustrated.

Chapter 3 is devoted to the presentation of surrogate scalar functions and scalarizing techniques, namely selecting one objective function to be optimized considering the other objectives as constraints, optimizing a weighted sum of the objective functions, and minimizing a distance function to a reference point using different metrics. The role of preference information in scalarizing techniques is also discussed. A special emphasis is placed on the multiobjective simplex tableau and the decomposition of the parametric (weight) space into indifference regions.

Chapter 4 deals with interactive methods in MOLP. The interactive methods STEM, Zionts and Wallenius's method, TRIMAP, Interval Criterion Weights (ICW) and Pareto Race are thoroughly described and illustrated using examples. These interactive methods are representative of different solution computation techniques, reduction of the scope of the search and interaction schemes. These methods were integrated in the TOMMIX method base developed in the early 1990s (Antunes et al. 1992), which enabled the information transfer among them. Some of the TOMMIX features are now included in the interactive software iMOLPe.

Chapter 5 presents the interactive MOLP explorer (iMOLPe) software, which is a computational package to tackle MOLP problems developed by the authors and accompanies this book. iMOLPe has been mainly designed for teaching and decision support purposes in MOLP problems. The aim is to offer students an intuitive environment as the entrance door to multiobjective optimization in which the main theoretical and methodological concepts can be apprehended through experimentation, thus enabling them to learn at their own pace.

Chapter 6 presents multiobjective integer (MOILP) and mixed-integer linear programming (MOMILP) problems, which are more difficult to tackle even having linear objective functions and constraints. The feasible set is no longer convex and in many cases these problems cannot be handled by adaptations of MOLP methods to deal with integer variables. Generating methods and scalarizing processes are reviewed and interactive methods devoted to MOILP/MOMILP problems are outlined. An interactive reference point method developed by the authors using branch-and-bound to perform directional searches in MOMILP is presented and illustrated using a computational implementation.

All chapters end with a comprehensive set of proposed exercises to extend the training provided by the illustrative examples in the text.

The interactive MOLP explorer (iMOLPe) software, as well as the computational implementation to MOMILP problems, can be freely downloaded at: www.inescc.pt/software.

# Chapter 2
# Formulations and Definitions

## 2.1 Introduction

Multiobjective Programming (MOP) may be faced as the extension of classical single objective programming to the cases in which more than one objective function is explicitly considered in mathematical optimization models. However, if these functions are conflicting, a paradigm change is at stake. The concept of optimal solution no longer makes sense since, in general, there is no feasible solution that simultaneously optimizes all objective functions. Single objective programming follows the optimality paradigm, that is, there is a complete comparability between pairs of feasible alternatives and transitivity applies. This is a mathematically well-formulated problem, since we possess enough mathematical tools to solve the three fundamental questions of analysis: existence, unicity and construction of the solution. When more than one objective function is considered these properties are no longer valid.

Zeleny (1982) shows in a very expressive manner the essential differences between single and multiple objective optimization models. Consider, following a similar reasoning, a bag of oranges, where the objective is to select firstly the biggest and then simultaneously the biggest and the sweetest orange. Although this is not a MOP example, its analysis is very suggestive and the conclusions are valid also in MOP. The selection of the biggest orange is akin to an optimization problem, hence purely technical. It is all about measuring and ordering the oranges. The same process arises in any optimization problem regarding feasible solutions. In the second case, if the biggest orange is not the sweetest one then there is no optimal solution to the problem. The selection of an orange forces a subjective compromise between the two objectives, size and sweetness, and thus problem solving is not limited to purely technical issues. However, there is a subset of oranges where the compromise solution must belong, the nondominated solution set, consisting of the oranges for which there is no other orange that is simultaneously bigger and sweeter.



C. Henggeler Antunes et al., *Multiobjective Linear and Integer Programming*,
EURO Advanced Tutorials on Operational Research,
DOI 10.1007/978-3-319-28746-1_2

Let us begin to present MOP models when all the $p$ objective functions and $m$ constraints are linear ones and decision variables are continuous. The multiobjective linear programming (MOLP) problem can be stated as:

$$\begin{aligned}
&\max\ z_1 = f_1(\mathbf{x}) = \mathbf{c}_1\mathbf{x} = \sum_{j=1}^{n} c_{1j}x_j \\
&\ldots \\
&\max\ z_p = f_p(\mathbf{x}) = \mathbf{c}_p\mathbf{x} = \sum_{j=1}^{n} c_{pj}x_j \\
&\text{s. t.}\ \sum_{j=1}^{n} a_{ij}x_j = b_i \qquad i = 1, \ldots, m \\
&\qquad x_j \geq 0 \qquad j = 1, \ldots, n
\end{aligned} \tag{2.1}$$

or

$$\left.\begin{aligned}
&\max\ z_1 = f_1(\mathbf{x}) = \mathbf{c}_1\mathbf{x} \\
&\qquad \ldots \\
&\max\ z_p = f_p(\mathbf{x}) = \mathbf{c}_p\mathbf{x}
\end{aligned}\right\} \ \text{"Max" } \mathbf{z} = \mathbf{f}(\mathbf{x}) = \mathbf{C}\mathbf{x}$$

$$\text{s. t.}\ \ \mathbf{x} \in X = \{\mathbf{x} \in \mathbb{R}^n : \mathbf{A}\mathbf{x} = \mathbf{b}, \mathbf{x} \geq \mathbf{0}\}.$$

$\mathbf{C} = \begin{bmatrix} \mathbf{c}_1 \\ \vdots \\ \mathbf{c}_p \end{bmatrix}$ and $\mathbf{c}_1, \cdots, \mathbf{c}_p$ are row vectors, $1 \times n$; $\mathbf{c}_k$ corresponds to the coefficients of objective function $k$ ($k = 1, \ldots p$). $\mathbf{A}$ is the technological coefficients matrix ($m \times n$), and all constraints were converted into equalities, with the introduction of auxiliary slack or surplus variables. $\mathbf{b} \in \mathbb{R}^m$ is the right hand side vector (in general, representing available resources for $\leq$ constraints and requirements for $\geq$ constraints). It is assumed that the feasible region $X$ is non-empty and a maximum exists in the feasible region for all objective functions being maximized. Without loss of generality, just to facilitate notation, we consider the maximization of each function[1] $f_k(\mathbf{x})$, $k = 1, \ldots, p$.

While in the optimization of a single objective function the feasible region in the decision space $\mathbf{x} \in X$ is mapped onto $\mathbb{R}$, in the multiobjective case the decision space is mapped onto a $p$-dimensional space $Z = \{\mathbf{z} = \mathbf{f}(\mathbf{x}) \in \mathbb{R}^p: \mathbf{x} \in X\}$, which is called *objective function space* or *criterion space*. In this space each potential solution $\mathbf{x} \in X$ is represented by a vector $\mathbf{z} = (z_1, z_2, \ldots, z_p) = \mathbf{f}(\mathbf{x}) = (f_1(\mathbf{x}), f_2(\mathbf{x}), \ldots, f_p(\mathbf{x}))$, the components of which are the values of each objective function for solution $\mathbf{x}$ of the feasible region (Fig. 2.1).

[1] Nothing would change, in substance, when considering minimization problems or cases where some objective functions should be maximized and others minimized. In this case the original problem is transformed into another problem where all objective functions are maximized (or minimized), by multiplying the minimizing objective functions by $-1$ (or the opposite).

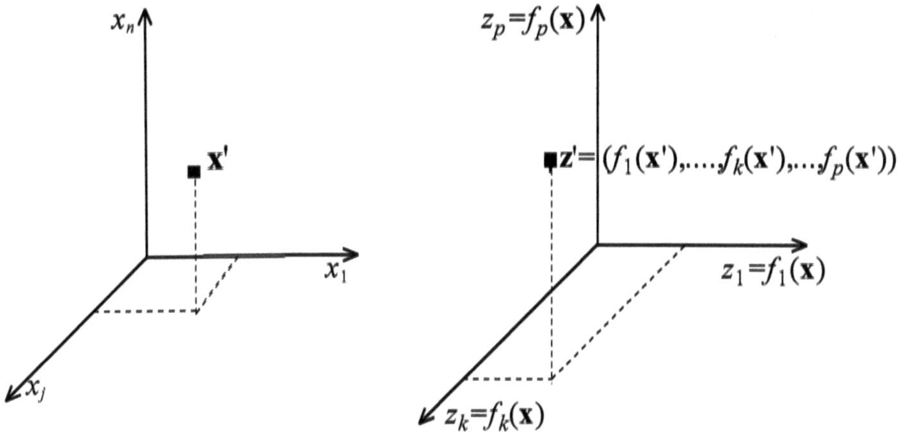


**Fig. 2.1** The decision space and the objective function space

Since, in general, the objective functions are conflicting, there is no feasible solution $\mathbf{x}\in X$ that simultaneously optimizes all objective functions. Therefore, from an operational perspective, the "Max" operation in (2.1) represents the computation of compromise solutions for which it is not possible to improve an objective function value without accepting to worsen, at least, the value of another objective function. These privileged solutions are called *efficient*, *nondominated*, *noninferior* or *Pareto optimal* solutions, and they are of interest to be the possible outcome of a decision process based on the vector optimization model.

## 2.2 Fundamental Concepts

In order to introduce the major fundamental concepts in a MOLP setting, let us consider a simple production planning problem with a single objective function. A small workshop manufactures two different products, I and II. The production of these products requires the use of three different types of machines—A, B and C. Each unit of product I requires 1 h processing in machine A, 2 h in machine B, and 2 h in machine C. Each unit of product II requires 1 h processing in machine A, 1 h in machine B, and 5 h in machine C. The workshop has a weekly maximum usage of 50 h of machine A, 80 h of machine B, and 220 h of machine C. The profit associated with of 1 unit of each product is 25 monetary units (m.u.) for product I and 20 m.u. for product II. It is assumed that all the production is sold.

Determining the optimal weekly production mix to maximize profit would amount to developing the following LP model (*Problem* 1), in which the decision variables $x_1$ and $x_2$ represent the number of units of product I and II, respectively, to be manufactured weekly. The constraints (of type $\leq$) refer to the availability of each type of machine and the objective function operationalizes the measure of performance of the system (to maximize the profit).

**Problem 1**

$$\max\ z = f(\mathbf{x}) = 25x_1 + 20x_2 \text{ (m.u.)}$$
$$\begin{aligned} \text{s.t.}\quad x_1 + x_2 &\leq 50 && \text{– h/week in machines of type A : (1)} \\ 2x_1 + x_2 &\leq 80 && \text{– h/week in machines of type B : (2)} \\ 2x_1 + 5x_2 &\leq 220 && \text{– h/week in machines of type C : (3)} \\ x_1, x_2 &\geq 0 \end{aligned}$$

The problem can be solved graphically (Fig. 2.2). The shadowed area displays the region of combinations $(x_1, x_2)$ that satisfy all the constraints, i.e., the feasible region $X$. For a specific value of $z$, $z = 25\ x_1 + 20\ x_2$ represents a level line of the objective function, which is normal to the gradient of the objective function (the vector (25, 20)). Since the objective function is increasing in the direction of its gradient, the optimal solution is found at the corner point (vertex) of the feasible region given by the intersection of $x_1 + x_2 = 50$ and $2x_1 + x_2 = 80$, i.e., where both constraints (1) and (2) are satisfied as equality, which means that the corresponding resources are fully consumed. Thus, the optimal solution is $x_1^* = 30$, $x_2^* = 20$ offering a profit of $z^* = 25 \times 30 + 20 \times 20 = 1150$.

Let us suppose that, in addition to profit, products I and II lead to other benefits (to be maximized) or losses (to be minimized) that cannot be aggregated in the profit objective function because they cannot be monetarily measured, e.g., maximizing

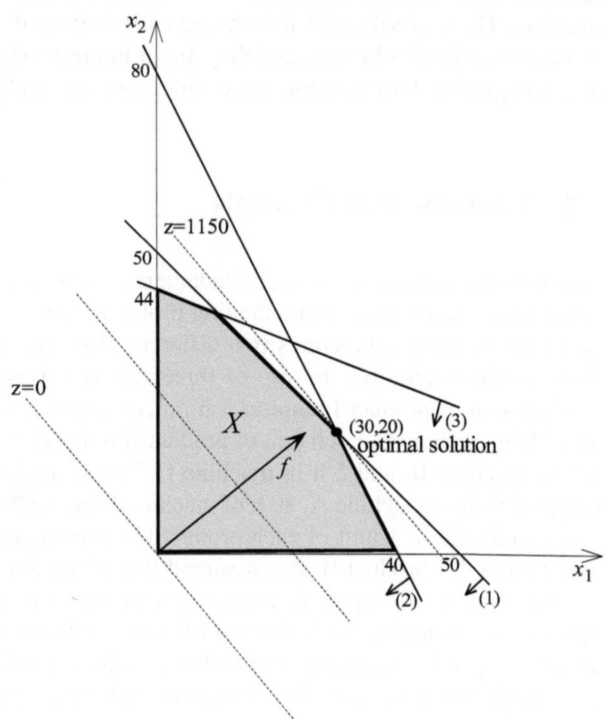


**Fig. 2.2** Graphical representation of the LP *Problem* 1

the conservation state of the machines or minimizing the environmental impact of the wastes generated by the production process. For instance, a new objective function is considered measuring the reliability of the production system, which depends on the quantities manufactured of products I and II: max $z_2 = x_1 + 8x_2$. This bi-objective problem is formulated in *Problem* 2.

**Problem 2**

$$\begin{aligned}
\max\ z_1 &= f_1(\mathbf{x}) = 25x_1 + 20x_2 && \text{(profit)}\\
\max\ z_2 &= f_2(\mathbf{x}) = x_1 + 8x_2 && \text{(reliability of the production system)}\\
\text{s.t.}\quad & x_1 + x_2 \leq 50\\
& 2x_1 + x_2 \leq 80\\
& 2x_1 + 5x_2 \leq 220\\
& x_1, x_2 \geq 0
\end{aligned}$$

As can be seen in Fig. 2.3, there is no feasible solution that simultaneously optimizes the two objective functions of *Problem* 2. Objective function $f_1(\mathbf{x})$ is optimized in solution $\mathbf{x} = (30, 20)$, point P, where $z_1 = 1150$ and $z_2 = 190$; objective function $f_2(\mathbf{x})$ is optimized in solution $\mathbf{x} = (0, 44)$, point R, where $z_1 = 880$ and $z_2 = 352$. The vertex solution identified in Fig. 2.3 by Q, $\mathbf{x} = (10, 40)$, where $z_1 = 1050$ and $z_2 = 330$, is an intermediate solution between P and R. Q is better than R in $f_1(\mathbf{x})$ (and worse in $f_2(\mathbf{x})$) and Q is better than P in $f_2(\mathbf{x})$ (and worse in $f_1(\mathbf{x})$).

Solutions P, Q and R are called *efficient*, because there is no other feasible solution that performs equal or better for both objective functions, and strictly better for at least one of those objective functions. The same happens with any solution on the edges [PQ] and [QR].

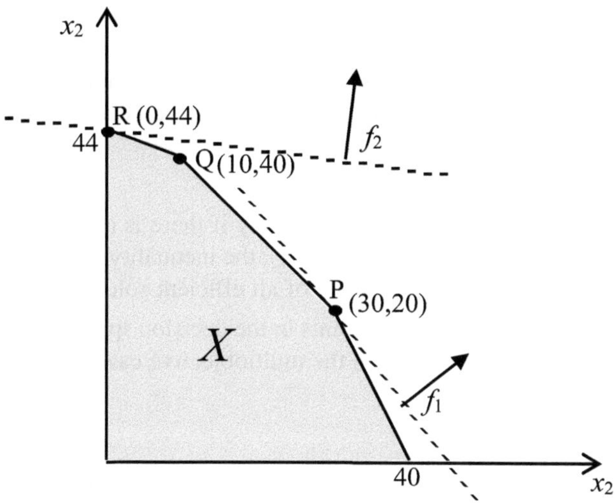


**Fig. 2.3** Graphical representation of the bi-objective LP *Problem* 2 in the decision variable space

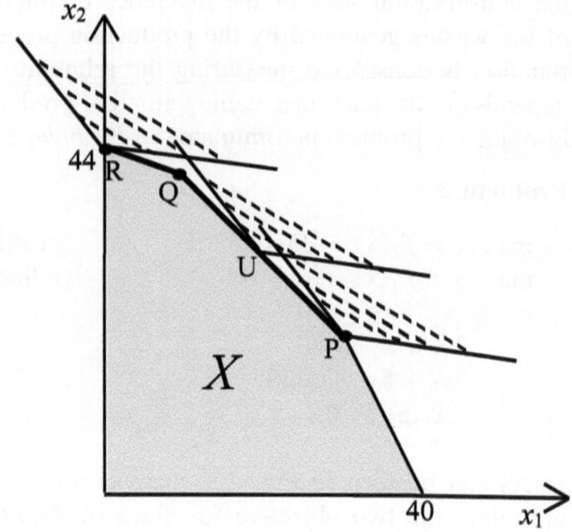


**Fig. 2.4** Identification of efficient solutions to the bi-objective LP *Problem* 2 using dominance cones

Figure 2.4 shows, for vertex solutions P and R, as well as point U on the edge [PQ], the cones associated with the objective function gradients, which can be designated by *dominance cones*, where better solutions for both objective functions would be located. Besides points P, R and U, there is no intersection of these cones with the feasible region. Hence, all solutions belonging to [PQ]∪[QR] (including the vertices) are called *efficient*, because they are not dominated by any other feasible solution. Note that this does not happen with any other solution in the feasible region not belonging to the frontier [PQ]∪[QR].

Figure 2.5 illustrates the regions of solutions that *dominate* two particular solutions, V and T. These solutions are *not efficient* because there are other feasible solutions that improve simultaneously both objective functions. The solutions that dominate V and T lay on the intersection of the respective dominance cones emanating from V and T with the feasible region.

**Definition of efficient solution**[2]

A solution $\mathbf{x}^1 \in X$ is called *efficient* if and only if there is no other solution $\mathbf{x} \in X$ such that $f_k(\mathbf{x}) \geq f_k(\mathbf{x}^1)$ for all $k$ ($k = 1, \ldots, p$), the inequality being strict for at least one $k$ ($f_k(\mathbf{x}) > f_k(\mathbf{x}^1)$). $X_E$ denotes the set of all efficient solutions.

While in single objective LP the points in the decision space have an image in $\mathbb{R}$ mapped by the objective function, in the multiobjective case the images are in $\mathbb{R}^p$;

[2] The mathematical definition of efficient solution can be done in several ways. For example, Yu (1974) presents it in terms of extreme point cones; Lin (1976) uses the notion of directional convexity and Payne et al. (1975) use a perturbation function similar to the one introduced by Geoffrion (1971) in another context.

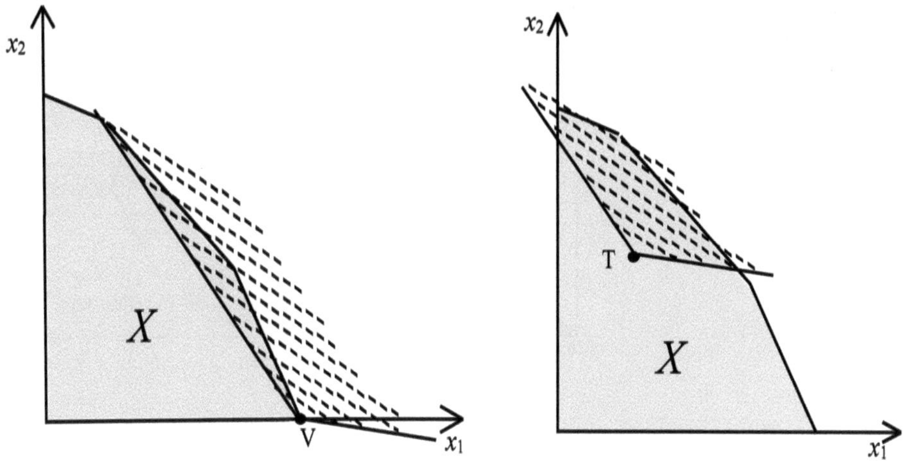


**Fig. 2.5** Examples of non-efficient solutions to the bi-objective *Problem* 2

that is, each solution $\mathbf{x}$ has a point $\mathbf{z} = (f_1(\mathbf{x}), \ldots, f_p(\mathbf{x}))$ as representation in the objective function space.

In *Problem* 2 the objective function space is two-dimensional and, hence, it is easy to visualize it graphically. In order to represent the feasible region of the problem in the objective function space *Z*, the images of the vertices in *X* are determined (which correspond to vertices in *Z*):

| | $\mathbf{x}$ | | | $\mathbf{z} = \mathbf{f}(\mathbf{x})$ |
|---|---|---|---|---|
| O | (0,0) | | O′ | (0,0) |
| V | (40,0) | ⇒ | V′ | (1000,40) |
| P | (30,20) | | P′ | (1150,190) |
| Q | (10,40) | | Q′ | (1050,330) |
| R | (0,44) | | R′ | (880,352) |

The considerations above about efficient solutions can be transposed into the objective function space (Fig. 2.6).

In general, while the designation of *efficient* solution is referred to points in the decision variable space, the designation of *nondominated* solution is used for points in the objective function space. When used in a generic way in this text, the designations of efficient and nondominated solution are used interchangeably.

**Definition of nondominated solution**

A point in the objective function space $\mathbf{z} = (f_1(\mathbf{x}), f_2(\mathbf{x}), \ldots, f_p(\mathbf{x})) \in Z$ is called *nondominated* if and only if $\mathbf{x}$ is efficient. $Z_E = \{\mathbf{z} = \mathbf{f}(\mathbf{x}) \in Z\text{: } \mathbf{x} \in X_E\}$.

In addition to the concept of *(strictly) efficient (nondominated)* solution, there is a "relaxed" concept of *weakly efficient solution* (*weakly nondominated solution*), i.e., a feasible solution is said weakly efficient/nondominated if and only if there is no other feasible solution that strictly improves the value of all objective functions.

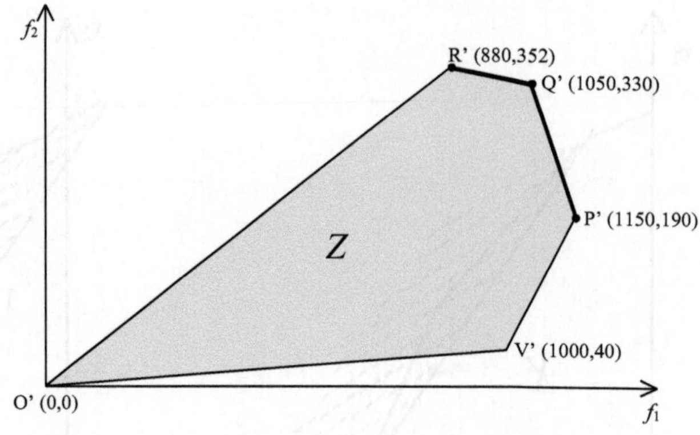


**Fig. 2.6** Feasible region of *Problem* 2 in the objective function space

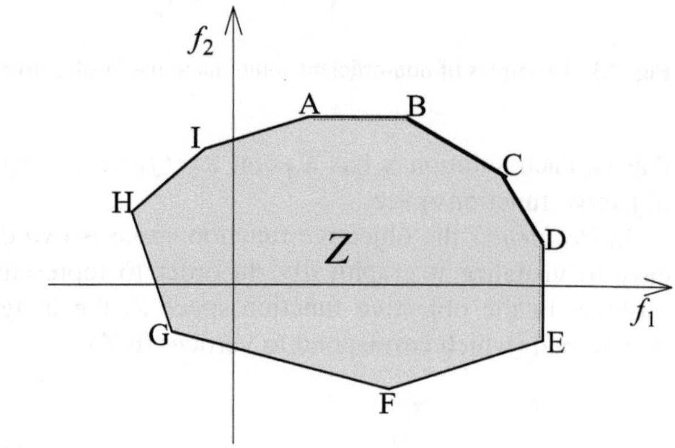


**Fig. 2.7** Illustration of strictly and weakly nondominated solutions

**Definition of weakly efficient/nondominated solution**

A solution $\mathbf{x}^1 \in X$ is called *weakly efficient* if and only if there is no other solution $\mathbf{x} \in X$ such that $f_k(\mathbf{x}) > f_k(\mathbf{x}^1)$ for all $k$ ($k = 1, \ldots, p$). $X_{WE}$ denotes the set of weakly efficient solutions.

A point in the objective function space $\mathbf{z} = (f_1(\mathbf{x}), f_2(\mathbf{x}), \ldots, f_p(\mathbf{x})) \in Z$ is called *weakly nondominated* if and only if $\mathbf{x} \in X_{WE}$, that is $Z_{WE} = \{\mathbf{z} = \mathbf{f}(\mathbf{x}) \in Z \colon \mathbf{x} \in X_{FE}\}$.

Note that, by definition, the set of weakly efficient solutions includes the strictly efficient solutions. However, for practical reasons, when weakly efficient solutions are mentioned in this text, the strictly efficient solutions are not being considered.

Figure 2.7 illustrates the concepts of weakly nondominated and strictly nondominated solutions with two objective functions to maximize: solutions on the edges [AB[ and [ED[ are just weakly nondominated, i.e., except the points B and D, while solutions on the edges [BC] and [CD] are (strictly) nondominated.

No weakly nondominated solutions exist in *Problem* 2 (Fig. 2.6). In order to illustrate this case, consider the temporary change of the second objective function

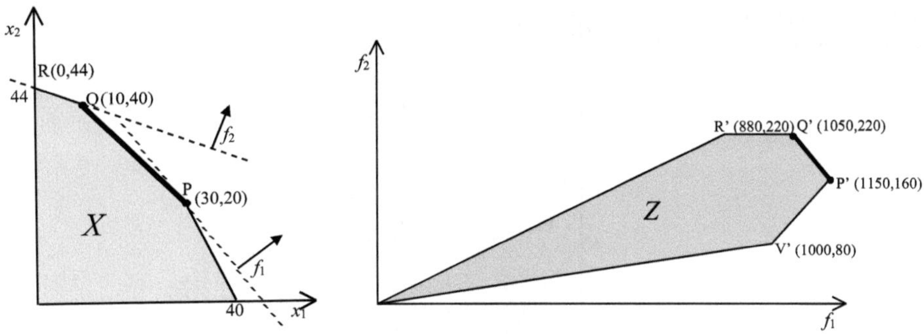


**Fig. 2.8** Weakly efficient/nondominated solutions (*Problem* 2 with $f_2(\mathbf{x}) = 2\,x_1 + 5\,x_2$)

to $f_2(\mathbf{x}) = 2\,x_1 + 5\,x_2$. The set of (strictly) efficient solutions is now reduced to the segment [PQ] in Fig. 2.8. All solutions on [RQ[ (note that Q is excluded) are weakly efficient because they are dominated by Q, which has equal $f_2(\mathbf{x})$ value and a better $f_1(\mathbf{x})$ value. Thus, there is no solution that simultaneously improves both objective functions regarding any solution on [RQ[.

Note that in problems with two objective functions weakly efficient/nondominated solutions can only occur when there are alternative optimal solutions of some objective function. In problems with $p \geq 3$ weakly efficient/nondominated solutions may also appear in other circumstances.

Therefore, in a multiobjective problem there is the need to select a compromise solution from the nondominated solution set, which entails a certain trade-off between the competing objective functions. A further illustration of the compromise involved when multiple objectives are at stake is made below, with three objective functions and two decision variables.

**Problem 3**

A family needs to define a diet of minimum cost which satisfies certain requirements of nutrients using basic foods I and II. The family wants to determine the intake of I and II in order to consume at least 200 units (u.) of vitamin A, 66 u. of vitamin B and 60 u. of vitamin C. Each 100 g of food I gives 5 u. of vitamin A, 3 u. of vitamin B and 1 u. of vitamin C, and costs 72 monetary units (m.u); each 100 g of food II gives 4 u. of vitamin A, 1 u. of vitamin B and 4 u. of vitamin C, and costs 35 m.u. Representing by $x_1$ and $x_2$ the consumption, in 100 g, of each food I and II, respectively, the problem formulation is:

$$
\begin{array}{lll}
\min \mathrm{z} = 72x_1 + 35x_2 \ (\mathrm{m.u}) & & \\
\mathrm{s.t.} \quad 5x_1 + 4x_2 \geq 200 & - \text{ requirement of vitamin A}: & (1) \\
\qquad\quad 3x_1 + x_2 \geq 66 & - \text{ requirement of vitamin B}: & (2) \\
\qquad\quad x_1 + 4x_2 \geq 60 & - \text{ requirement of vitamin C}: & (3) \\
\qquad\quad x_1, x_2 \geq 0 & &
\end{array}
$$

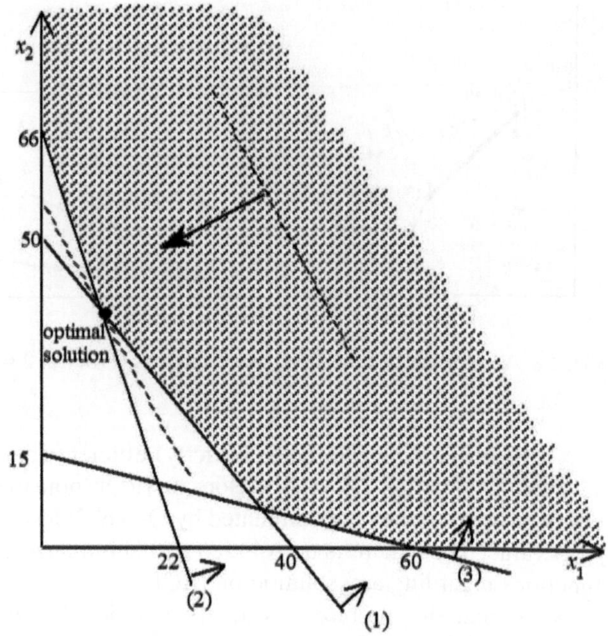


**Fig. 2.9** Graphical representation of *Problem* 3

The (unbounded) feasible region is displayed in Fig. 2.9. The optimal solution is $x_1^* = 9.143, x_2^* = 38.571$, i.e., the minimum cost diet establishes an intake of 914.3 g of food I and 3857.1 g of food II, amounting to $z^* = 2008.281$ m.u.

**Problem 4**

Let us suppose that the problem now is maximizing the intake of the three vitamins, A, B and C, subject to the cost of the diet not exceeding the minimum cost obtained in the previous formulation (2008.281 m.u.). The formulation of this problem is:

$$\begin{aligned}
&\max z_1 = f_1(\mathbf{x}) = 5x_1 + 4x_2 && \text{(vitamin A)}\\
&\max z_2 = f_2(\mathbf{x}) = 3x_1 + x_2 && \text{(vitamin B)}\\
&\max z_3 = f_3(\mathbf{x}) = x_1 + 4x_2 && \text{(vitamin C)}\\
&\text{s. t.} \quad 72x_1 + 35x_2 \le 2008.281 &&\\
&\qquad\qquad x_1, x_2 \ge 0 &&
\end{aligned}$$

The feasible region, in the decision space, and the gradients of the objective functions of this problem are depicted in Fig. 2.10.

All solutions on [PQ] are efficient to this problem (Fig. 2.10). Solution P, $\mathbf{x} = (27.893, 0)$, $\mathbf{z} = (139.465, 83.679, 27.893)$, provides a level of vitamin B above the previously required (with the maximum of vitamin B, $f_2(\mathbf{x})$), but lower levels of vitamins A and C. Solution Q, $\mathbf{x} = (0, 57.379)$, $\mathbf{z} = (229.516, 57.379, 229.516)$, offers high levels for vitamins A and C (the maxima of $f_1(\mathbf{x})$ and $f_3(\mathbf{x})$), although decreasing vitamin B below the minimum previously required. Any other solution

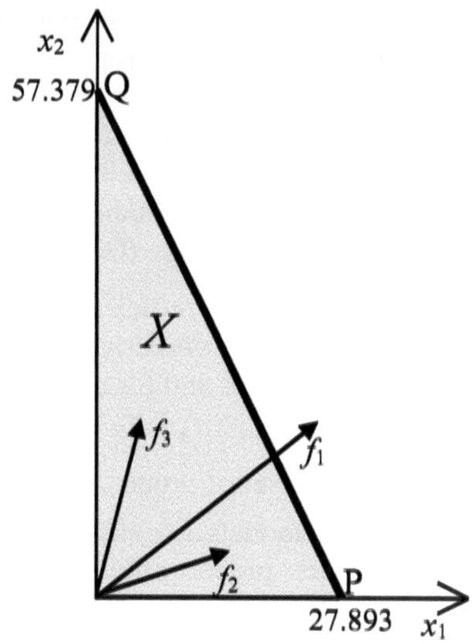


**Fig. 2.10** Feasible region to *Problem* 4 in the decision space

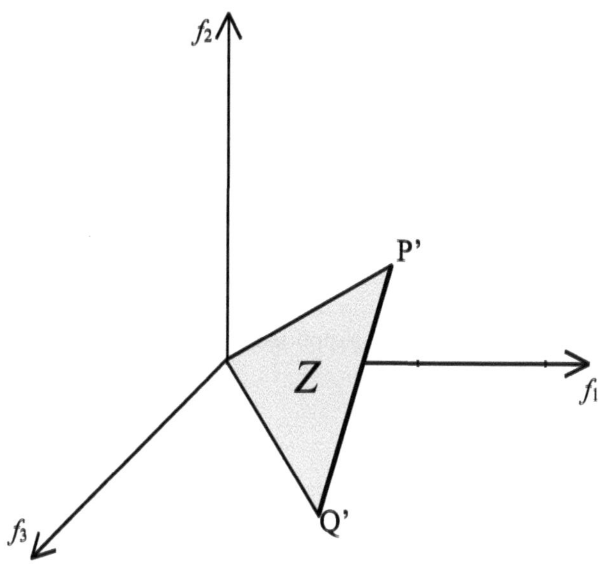


**Fig. 2.11** Feasible region to *Problem* 4 in the objective function space

on the edge that links P to Q is an intermediate compromise solution, including the optimal solution to *Problem* 3, which is $\mathbf{x} = (9.143, 38.571)$, $\mathbf{z} = (200, 66, 163.427)$.

The feasible region of *Problem* 4 in the objective function space is depicted in Fig. 2.11. All solutions on [P'Q'] are nondominated to this problem, i.e., they are the images of [PQ].

The consideration of integer (or just binary) variables in MOLP models as well as the existence of non-linearities in the objective functions and/or constraints require some additional concepts. New types of efficient/nondominated solutions arise in these problems: proper, improper, supported and unsupported solutions.

The definition of *proper efficient solution* embodies a more restrictive notion of efficient solution in order to exclude efficient solutions that display unbounded compromises between the objective functions, i.e., to exclude solutions in which the relation improvement/deterioration between the objective function values can be made arbitrarily large (Geoffrion 1968).

**Definition of proper efficient solution**

A solution $\mathbf{x}' \in X$ is properly efficient if it is efficient and a finite $\mathrm{M} > 0$ exists such that for each $\mathbf{x} \in X$ and for each objective function $f_k(\mathbf{x})$, $k = 1, \ldots, p$ with $f_k(\mathbf{x}) > f_k(\mathbf{x}')$, the relation $\frac{f_k(\mathbf{x}) - f_k(\mathbf{x}')}{f_j(\mathbf{x}') - f_j(\mathbf{x})} \leq \mathrm{M}$ is verified for some $j$ for which $f_j(\mathbf{x}) < f_j(\mathbf{x}')$. $X_{PE}$ represents the set of proper efficient solutions.

In MOLP models $X_{PE} \equiv X_E$. Also in integer LP and mixed integer LP all efficient solutions are proper efficient. However, in nonlinear multiobjective models improper efficient solutions may exist.

Figure 2.12 illustrates the concept of *improper efficient solution* in the objective function space for two nonlinear bi-objective problems with both functions to be maximized. In Fig. 2.12a the efficient solutions lie on the arcs AB and CD, excluding point D which is just weakly efficient. Solutions A, B and C are improper efficient solutions. In Fig. 2.12b, the entire frontier from A to C is efficient, passing through point B which is an improper efficient solution.

Another important issue is the distinction between supported and unsupported efficient solutions.

**Definition of unsupported efficient/nondominated solution**

A nondominated solution $\mathbf{z}' \in Z_E$ is unsupported if it is dominated by a (infeasible) convex combination of solutions belonging to $Z_E$. An unsupported nondominated solution $\mathbf{z}' = \mathbf{f}(\mathbf{x}')$ is the image of an unsupported efficient solution $\mathbf{x}'$. The other efficient solutions are supported.

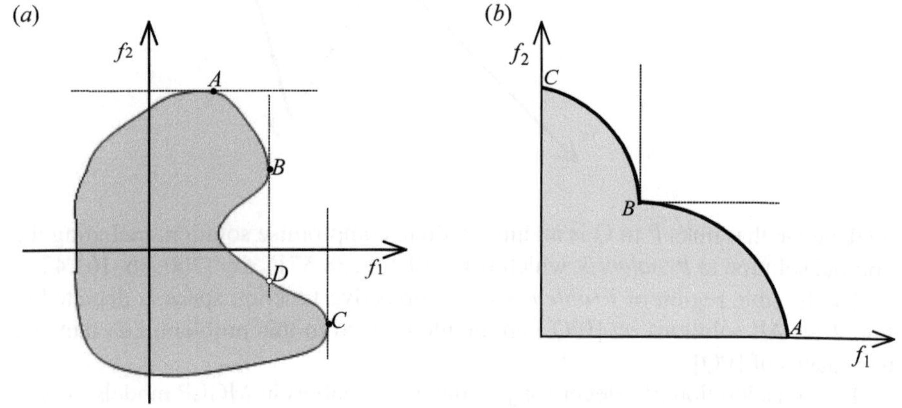


**Fig. 2.12** Improper efficient solutions in nonlinear problems ((b) in Steuer 1986)

In MOLP models all nondominated solutions are supported. However, in problems for which $Z$ is non-convex, unsupported nondominated solutions may exist. The cases of multiobjective integer and mixed integer LP as well as multiobjective nonlinear models are illustrated below.

Figure 2.13 illustrates the existence of unsupported nondominated solutions in a multiobjective integer problem, with both functions to be maximized. Points A, B and D are supported nondominated solutions, while C is an unsupported nondominated solution because it is dominated by some (infeasible) convex combinations of B and D, i.e., all convex combinations defined by the intersection of the dominance cone emanating from C with the segment connecting B and D. That is, C lies inside the convex hull defined by the supported solutions.

Figure 2.14 shows the nondominated frontier of a mixed-integer LP problem with two objective functions to be maximized. Solution D and all solutions on the

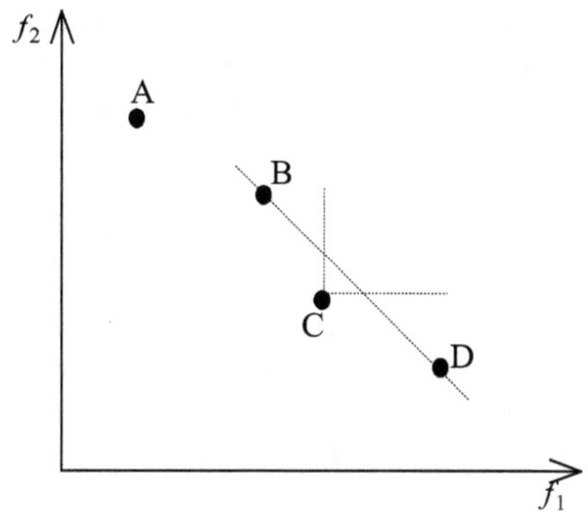


**Fig. 2.13** Supported and unsupported nondominated solutions in integer programming

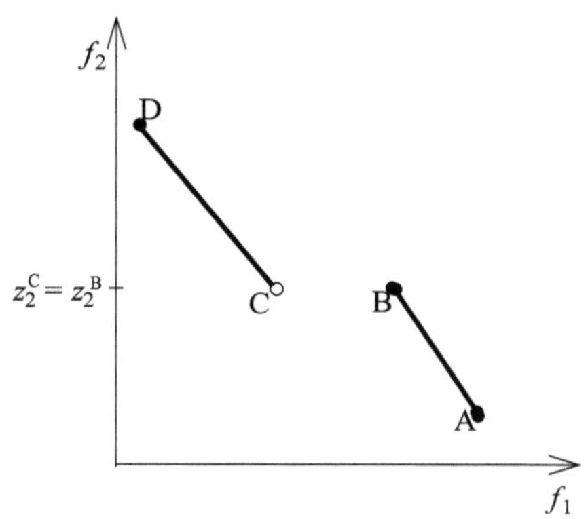


**Fig. 2.14** Unsupported nondominated solutions in a bi-objective mixed-integer problem

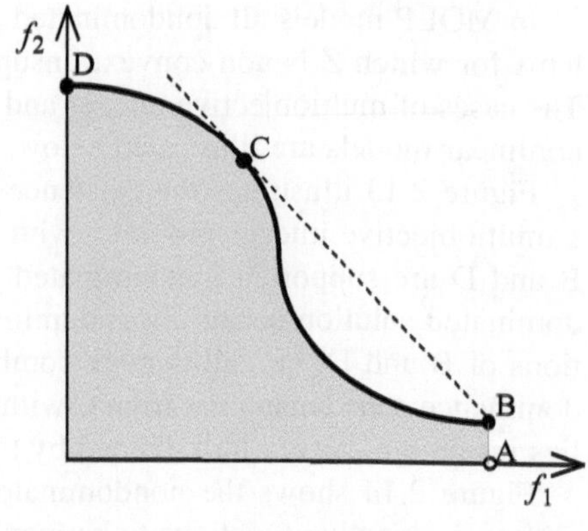


**Fig. 2.15** Unsupported nondominated solutions in a bi-objective nonlinear problem

segment [AB] are supported nondominated solutions. Solutions on the segment ]CD[ are unsupported nondominated solutions because they are nondominated although they are dominated by (infeasible) convex combinations of B and D. Solution C is weakly nondominated because there is no other solution strictly better than C in both objective functions, but it is dominated by solution B which has the same value of $f_2(\mathbf{x})$ and a better value of $f_1(\mathbf{x})$.

Figure 2.15 displays the feasible region in the objective space for a bi-objective nonlinear problem, in which the nondominated frontier is displayed in thick solid line (solutions from B to D). The solutions on the segment [AB[, i.e., excluding B, are weakly nondominated solutions. The solutions from B (excluding B) to C (excluding C) are unsupported nondominated solutions. Solution B and the solutions on the arc CD are supported nondominated solutions.

The *ideal solution* and the *pay-off table* are often used in MOP as auxiliary devices to identify the range of variation of the objective function values over the nondominated region and help to develop auxiliary problems to compute nondominated solutions.

### Ideal Solution

The ideal solution (or utopia point) $\mathbf{z}^* = \left(z_1^*, z_2^*, \ldots, z_p^*\right)$ is defined as the solution that would simultaneously optimize all objective functions. That is, the components of the ideal solution are the individual optimal values to each objective function in the feasible region. In general, the ideal solution does not belong to the feasible region (otherwise the problem would be trivial because all objective functions would have their optimum at the ideal solution), although each component $z_k^*$, $k = 1, \ldots, p$, is individually reachable.

The ideal solution is often used as the (unreachable) reference point in scalarizing functions aimed at determining compromise nondominated solutions by minimizing a distance function to the ideal solution. In Fig. 2.16 the ideal solution to *Problem* 2 is displayed.

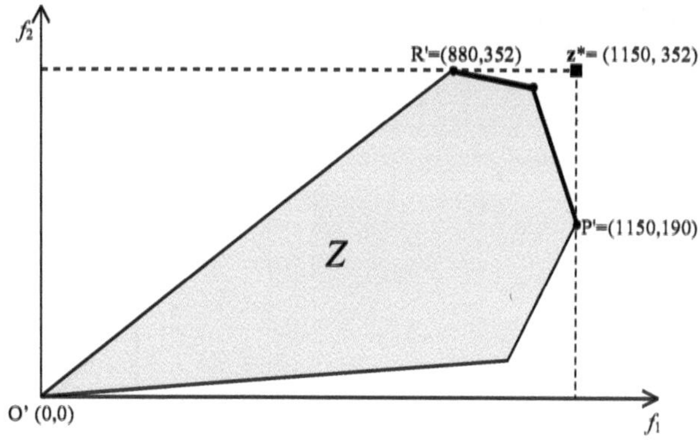


**Fig. 2.16** Ideal solution to *Problem* 2

Note that although the ideal solution $\mathbf{z}^*$ can be always defined in the objective function space, this is not always possible in the decision space: that is, $\mathbf{x}^*$, even infeasible, may not exist such that $\mathbf{z}^* = \mathbf{f}(\mathbf{x}^*)$.

**Pay-off Table**

The *pay-off table* provides the objective function values resulting from individually optimizing each objective function. This table allows having a global overview of the range of variation of the objective function values over the nondominated region.

| | $f_1(\mathbf{x})$ | $f_2(\mathbf{x})$ | ... | $f_k(\mathbf{x})$ | ... | $f_p(\mathbf{x})$ |
|---|---|---|---|---|---|---|
| $\mathbf{z}^1$ | $z_1^1 = z_1^*$ | $z_2^1$ | ... | $z_k^1$ | ... | $z_p^1$ |
| $\mathbf{z}^2$ | $z_1^2$ | $z_2^2 = z_2^*$ | ... | $z_k^2$ | ... | $z_p^2$ |
| ... | ... | ... | ... | ... | ... | ... |
| $\mathbf{z}^k$ | $z_1^k$ | $z_2^k$ | ... | $z_k^k = z_k^*$ | ... | $z_p^k$ |
| ... | ... | ... | ... | ... | ... | ... |
| $\mathbf{z}^p$ | $z_1^p$ | $z_2^p$ | ... | $z_k^p$ | ... | $z_p^p = z_p^*$ |

$z_k^i$ is the value of objective function $f_k(\mathbf{x})$ in the solution $i$. The solution $i$ is denoted by $\mathbf{z}^i$, $i = 1, \ldots, p$, and optimizes objective function $f_i(\mathbf{x})$ in the feasible region $X$. The optimal value of each objective function $f_k(\mathbf{x})$, $z_k^k = z_k^*$, can be found in the main diagonal of the table, therefore allowing to easily identify the ideal solution.

The nadir point gives the minimum (worst) objective function values over the set of all nondominated solutions. The ideal solution and the nadir point define the range of variation of each objective function over the nondominated region. Except for the case of two objective functions, the pay-off table does not offer, in general, the nadir point. The pay-off table displays the objective function values for the solutions that individually optimize each objective function; in general, the minimum of each objective function over the nondominated solution set is not attained in one of these solutions. Some approaches approximate the nadir point using the information in the pay-off table by representing the true nadir by a "convenient"

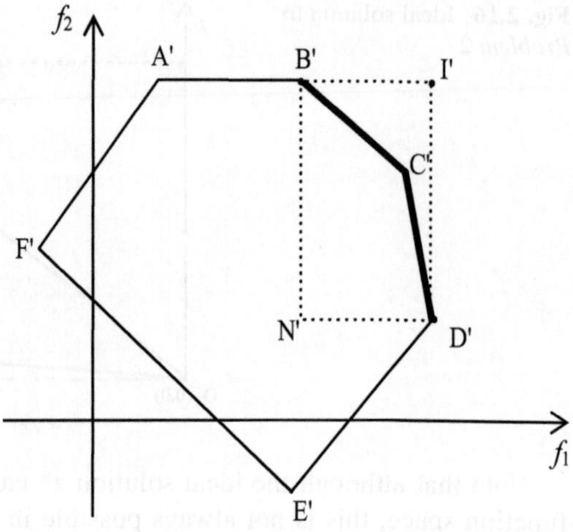


**Fig. 2.17** Ideal solution (I′) and nadir point (N′)

one directly obtained from the pay-off table: selecting in each column the worst value of the corresponding objective function, that is $n_k = min_{i=1,\ldots,p}\, z_k^i$, $k=1,\ldots,p$. This value $n_k$ is an approximation to the minimum of the objective function $f_k(\mathbf{x})$ in the nondominated region. In general, for $p>2$, there are nondominated solutions with worst values than the $n_k$ obtained from the *pay-off* table. The identification of the true nadir point is computationally very demanding for $p>2$. An exact method for this purpose in MOLP problems is described in Alves and Costa (2009).

Figure 2.17 represents the ideal solution I′ and the nadir point N′ for an example with two objective functions. Note that, for $p=2$, the nadir point obtained from the pay-off table always identifies the minimum of each objective function in the nondominated region, provided that all the solutions forming the pay-off table are nondominated and not just weakly nondominated.

Finally, note that the *pay-off* table for problems with $p>2$ may not be uniquely defined if there are alternative nondominated solutions that optimize any objective function. In this case, the approximation of the nadir point provided by the *pay-off* table is not unique although the ideal solution is always unique.

## 2.3 Proposed Exercises

1. Consider the problem:

$$\max \mathbf{f}(\mathbf{x}) = [f_1(x_1,x_2) = 5x_1 - 2x_2, f_2(x_1,x_2) = -x_1 + 4x_2]$$

s. t.

$$\begin{array}{ll} -x_1 + x_2 \le 3 & x_1 + x_2 \le 8 \\ x_1 \qquad \le 6 & x_2 \le 4 \\ x_1 \ge 0,\ x_2 \ge 0 & \end{array}$$

(a) Represent graphically the feasible region in the decision variable space and in the objective function space, identifying the set of efficient solutions and the set of nondominated solutions.
(b) Identify the ideal solution

2. Consider the problem:

$$\begin{array}{ll} \max & f_1(x_1,x_2) = -x_1 + 3x_2 \\ \max & f_2(x_1,x_2) = \ \ 3x_1 + 3x_2 \\ \max & f_3(x_1,x_2) = \ \ \ \ x_1 + 2x_2 \\ \text{s. t.} & x_2 \leq 4 \\ & x_1 + 2\ x_2 \leq 10 \\ & 2x_1 + \ \ x_2 \leq 10 \\ & x_1 \geq 0,\ x_2 \geq 0 \end{array}$$

(a) Represent graphically the feasible region and the gradients of the objective functions in the decision space.
(b) Compute the ideal solution.

3. Consider the problem with two objective functions:

$$\begin{array}{l} \min f_1(\mathbf{x}) = x_1 \\ \max f_2(\mathbf{x}) = x_2 \\ \text{s. t.} \quad \mathbf{x} \in X \end{array}$$

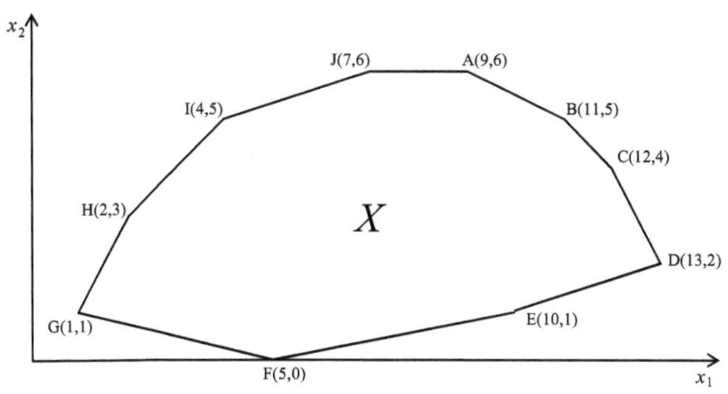


(a) Identify the set of efficient/nondominated solutions and the ideal solution.
(b) Identify the set of (just) weakly efficient/nondominated solutions.
(c) What will be the set of efficient solutions if the problem is:

$$\begin{array}{l} \max\ f_1(\mathbf{x}) = x_1 \\ \min\ f_2(\mathbf{x}) = x_1 \\ \text{s. t.} \quad \mathbf{x} \in X \end{array}$$

4.

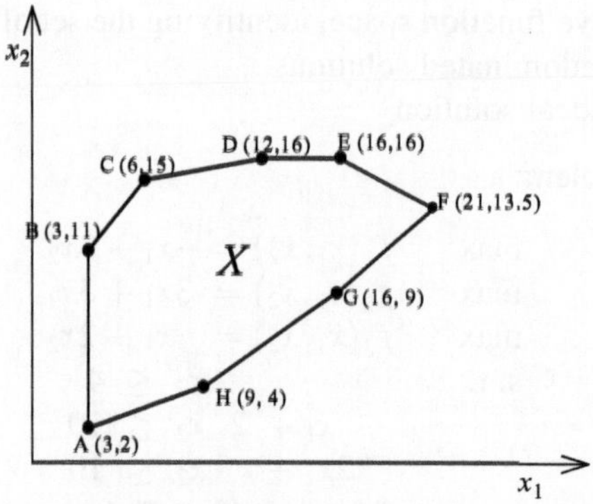


Consider the linear programming problem with two objective functions:

$$\begin{aligned} &\min\ f_1(\mathbf{x}) = 6x_1 - 5x_2 \\ &\max\ f_2(\mathbf{x}) = 4x_1 + 5x_2 \\ &\text{s. t.} \quad \mathbf{x} \in X \end{aligned}$$

(a) Sketch the feasible region in the objective function space, identifying the (strictly and weakly) nondominated region and the ideal solution.
(b) Identify graphically all feasible solutions that are dominated by point E.
(c) Identify graphically all feasible solutions that dominate point G.
(d) Identify the efficient/nondominated regions considering that

 (i) $x_1$ is integer
 (ii) both variables are integer.

# Chapter 3
# Surrogate Scalar Functions and Scalarizing Techniques

The most common procedure to compute efficient/nondominated solutions in MOP is using a scalarizing technique, which consists in transforming the original multiobjective problem into a single objective problem that may be solved repeatedly with different parameters. The functions employed in scalarizing techniques are called *surrogate scalar* functions or *scalarizing* functions. The optimal solution to these functions should be a nondominated solution to the multiobjective problem. These functions temporarily aggregate in a single dimension the $p$ objective functions of the original model and include parameters derived from the elicitation of the DM's preference information. Surrogate scalar functions should be able to generate nondominated solutions only, obtain any nondominated solution and be independent of dominated solutions. In addition, the computational effort involved in the optimization of surrogate scalar functions should not be too demanding (e.g., increasing too much the dimension of the surrogate problem or resorting to nonlinear scalarizing functions when all original objective functions are linear) and the preference information parameters should have a simple interpretation (i.e., not imposing an excessive cognitive burden on the DM). Surrogate scalar functions should not be understood as "true" analytical representations of the DM's preferences but rather as an operational means to transitorily aggregate the multiple objective functions and generate nondominated solutions to be proposed to the DM, which expectedly are in accordance with his/her (evolving) preferences.

Three main scalarizing techniques are generally used to compute nondominated solutions:

1. Selecting one of the $p$ objective functions to be optimized considering the other $p$-1 objectives as constraints by specifying the inferior (reservation) levels that the DM is willing to accept. This scalarization is usually called *e-constraint* technique.
2. Optimizing a weighted-sum of the $p$ objective functions by assigning weighting coefficients to them—*weighted-sum* technique.



C. Henggeler Antunes et al., *Multiobjective Linear and Integer Programming*,
EURO Advanced Tutorials on Operational Research,
DOI 10.1007/978-3-319-28746-1_3

3. Minimizing a distance function to a reference point (e.g., the ideal solution), the components of which are aspiration levels the DM would like to attain for each objective function. If the reference point is not reachable the closest solution according to a given metric is computed, usually the Manhattan metric (i.e., minimizing the sum of the differences in all objectives) or the Chebyshev metric (i.e., minimizing the maximum difference in all objectives) possibly considering weights, i.e., the differences are not equally valued for all objectives. However, the reference point may also be a point representing attainable outcomes. In this case, the surrogate scalar function is referred to as an *achievement* scalarizing function, as it aims to reach or surpass the reference point. These approaches are commonly referred to as *reference point* techniques.

The techniques will be presented for MOLP problems (cf. formulation (2.1) in Chap. 2) and then extended to integer and nonlinear cases. The theorems underlying the techniques for computing nondominated solutions are just sufficient conditions for efficiency. When these conditions are also necessary, then the corresponding technique guarantees the possibility to compute all nondominated solutions. Although this section pays special attention to sufficient conditions, necessary conditions shall not be forgotten since it is important to know the conditions in which all the nondominated solutions can be obtained using a given scalarizing technique.

## 3.1 Optimizing One of the Objective Functions and Transforming the Remaining p−1 into Constraints

**Proposition 1**
If $\mathbf{x}^1$ is the single optimal solution, for some $i$, to the problem

$$\begin{aligned} &\max f_i(\mathbf{x}) \\ &\text{s.t. } \mathbf{x} \in X \\ &\quad f_k(\mathbf{x}) \geq \mathrm{e}_k \qquad k = 1, \ \ldots, i-1, i+1, \ \ldots p \end{aligned} \tag{3.1}$$

then $\mathbf{x}^1$ is an efficient solution to the multiobjective problem.

If, in Proposition 1, the condition of a single optimal solution had not been imposed, weakly efficient solutions could be obtained. This issue could be overcome by replacing the function $f_i(\mathbf{x})$ by $f_i(\mathbf{x}) + \sum_{k \neq i} \rho_k f_k(\mathbf{x})$, with $\rho_k > 0$ small positive scalars.

The validity of this proposition assumes that the reduced feasible region is not empty, which may occur whenever the lower bounds $\mathrm{e}_k$ set on the $p$-1 objective functions that are transformed into constraints are too stringent.

The truthfulness of Proposition 1 is easily shown. Suppose that $\mathbf{x}^1$ is not efficient. Then, by definition of efficient solution, there is an $\mathbf{x}^2 \in X$ such that $f_k(\mathbf{x}^2) \geq f_k(\mathbf{x}^1)$ for all $k$ ($k = 1, \ldots, p$), and the inequality $f_k(\mathbf{x}^2) > f_k(\mathbf{x}^1)$ holds for at least one $k$. In these

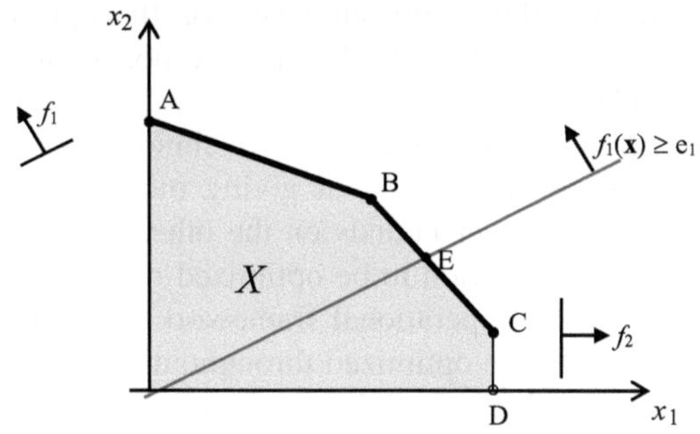


**Fig. 3.1** Computing an efficient solution by optimizing one of the objective functions and transforming the remaining $p-1$ into constraints

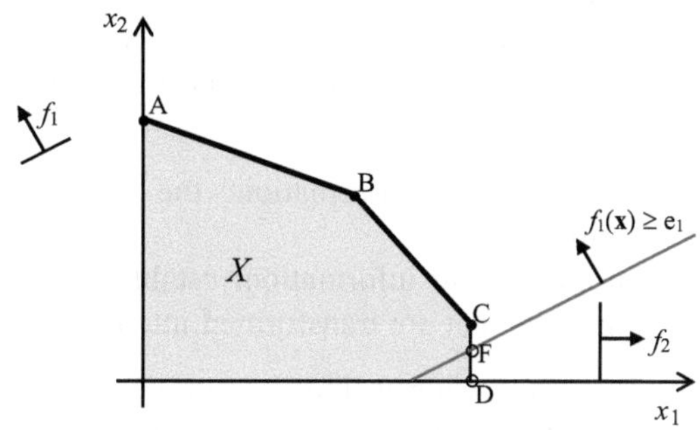


**Fig. 3.2** Computation of weakly efficient solutions

circumstances, $f_k(\mathbf{x}^2) \geq e_k$, for $k = 1, \ldots, i-1, i+1, \ldots, p$. Hence, $f_i(\mathbf{x}^2) \geq f_i(\mathbf{x}^1)$ in problem (3.1) which contradicts the hypothesis of $\mathbf{x}^1$ being the single optimal solution. Thus, $\mathbf{x}^1$ must be efficient.

This computation procedure is illustrated in a bi-objective LP model (Fig. 3.1), where $f_1(\mathbf{x})$ and $f_2(\mathbf{x})$ are maximized. The efficient frontier of the feasible region $X$ is composed by the solutions on edges [AB] and [BC]. Imposing the additional constraint $f_1(\mathbf{x}) \geq e_1$ and optimizing $f_2(\mathbf{x})$ then the efficient solution E is obtained. Note that E is not a vertex of the feasible region to the original problem.

The condition established in Proposition 1 is not a necessary condition for obtaining efficient solutions. In fact, efficient solutions can be obtained without having a single optimum to the scalarizing problem (3.1). In this case, not all solutions obtained are guaranteed to be efficient. If the optimum is not imposed to be unique, then a necessary and sufficient condition for obtaining at least weakly efficient solutions is achieved.

Figure 3.2 illustrates this issue: imposing $f_1(\mathbf{x}) \geq e_1$ and optimizing $f_2(\mathbf{x})$ then the edge [CF] is optimal but only point C is (strictly) efficient. Solutions on the edge [CF], except C, are just weakly efficient solutions.

An additional interest of this scalarizing technique is that the dual variable associated with the constraint corresponding to objective function $f_k(\mathbf{x})$ can be interpreted as a local trade-off rate between objectives $f_i(\mathbf{x})$ and $f_k(\mathbf{x})$ at the optimal solution of the scalar problem (3.1). The interpretation and use of this information should be done with care whenever this optimal solution is a degenerate one, since it this case the trade-off rates are not unique (i.e., alternative optima to the dual problem exist).

Although this scalarizing technique is simple to be understood by the DM, capturing the attitude of giving more importance to an objective function and accepting lower bounds on the other objective function values, the choice of the objective function to be optimized may reveal to be difficult in several problems. Also, in the operational framework of a particular method, setting the objective function to be optimized throughout the solution computation process may render the method less flexible and the results too dependent on the function selected.

Solving problem (3.1) enables to obtain all nondominated solutions, i.e., solutions lying on edges or faces (of any dimension) and vertices of the feasible region of the original multiobjective problem.

The preference information associated with this scalarizing technique consists in:

- inter-objective information: the selection of the objective function to be optimized;
- intra-objective information: establishing lower bounds on the other objective functions that are transformed into constraints.

This scalarizing technique can also be used in multiobjective integer, mixed-integer or nonlinear optimization, thus enabling to obtain any type of efficient/nondominated solution to these problems.

Figure 3.3 illustrates examples of bi-objective (a) integer and (b) nonlinear problems, in which $f_1(\mathbf{x})$ is optimized and $f_2(\mathbf{x})$ is considered as an additional constraint. In (*a*) unsupported nondominated solution C is obtained, and in (*b*) improper solution B is obtained.

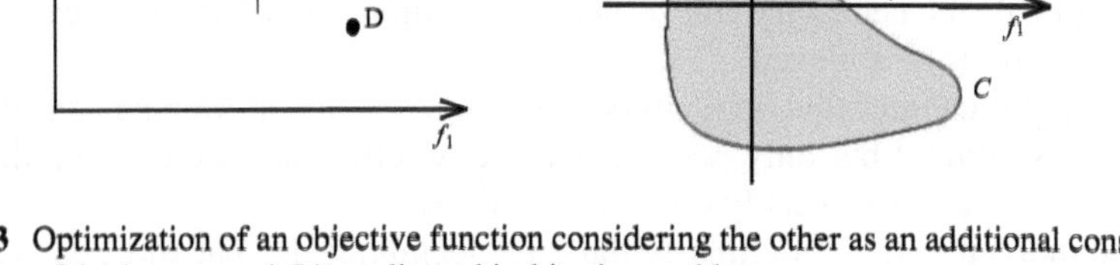


**Fig. 3.3** Optimization of an objective function considering the other as an additional constraint in the cases of (**a**) integer and (**b**) nonlinear bi-objective problems

## 3.2 Optimizing a Weighted-Sum of the Objective Functions

The process of computation of efficient/nondominated solutions more utilized consists in solving a scalar problem in which the objective function is a weighted-sum of the $p$ original objective functions with positive weights $\lambda_k$:

$$\begin{aligned} \max\ z_\lambda &= \sum_{k=1}^{p} \lambda_k f_k(\mathbf{x}) \\ \text{s.t.}\quad \mathbf{x} &\in X \end{aligned} \tag{3.2}$$

**Proposition 2**

If $\mathbf{x^1} \in X$ is a solution to the problem $\max\limits_{\mathbf{x}\in X} \sum_{k=1}^{p} \lambda_k f_k(\mathbf{x})$ for $\boldsymbol{\lambda} = (\lambda_1, \cdots, \lambda_p)$, where $\lambda_k > 0$, $k = 1, \ldots, p$, and $\sum_{k=1}^{p} \lambda_k = 1$, then $\mathbf{x^1}$ is an efficient solution to the multiobjective problem.

The truthfulness of Proposition 2 can be shown as follows. Suppose that $\mathbf{x^1}$ is not efficient. Then, there is an $\mathbf{x^2} \in X$ such that $f_k(\mathbf{x}^2) \geq f_k(\mathbf{x}^1)$, $k = 1, \ldots, p$, and the inequality is strict for at least one $k$. But $\mathbf{x^1}$ was obtained by optimizing a weighted-sum objective function with strictly positive weights; then $\sum_{k=1}^{p} \lambda_k f_k(\mathbf{x}^2) > \sum_{k=1}^{p} \lambda_k f_k(\mathbf{x}^1)$, which contradicts the hypothesis that $\mathbf{x^1}$ maximizes the weighted-sum objective function.

This computation procedure in MOLP is illustrated in Fig. 3.4. This figure also shows two weighted-sum objective functions, considering very different weight vectors (whose gradients are given by $f_{\lambda'}$ and $f_{\lambda''}$), can lead to the computation of the same efficient solution (point A).

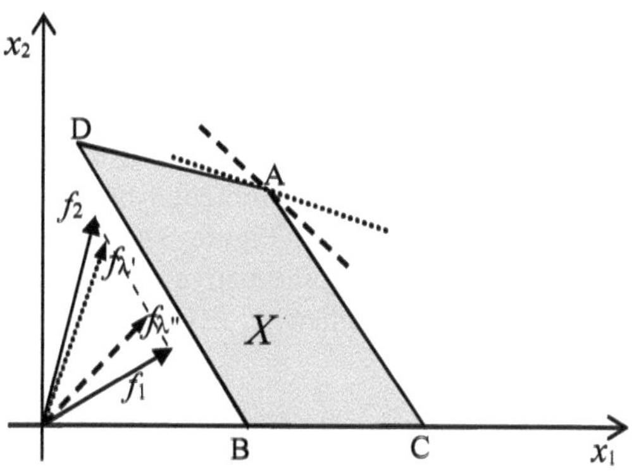


**Fig. 3.4** Optimizing weighted-sums of the objective functions

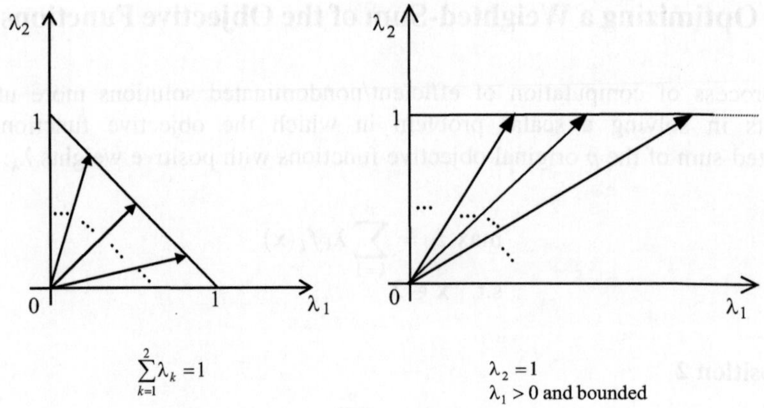


**Fig. 3.5** Weight normalization

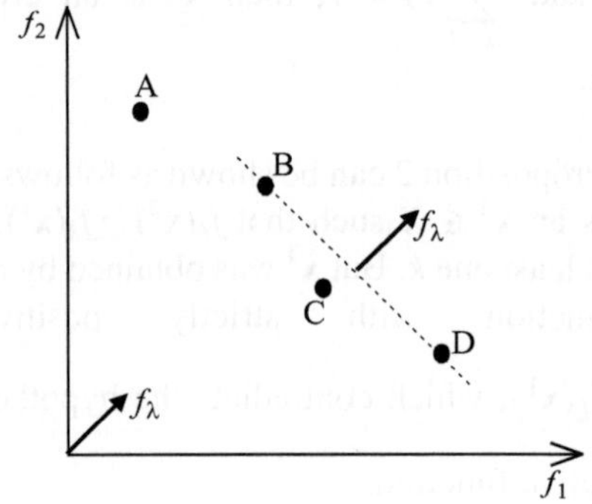


**Fig. 3.6** Optimization of a weighted-sum of the objective functions in integer programming

The weight normalization used in Proposition 2, $\sum_{k=1}^{p} \lambda_k = 1$, can be replaced by $\lambda_i = 1$, for a given $i$, $1 \leq i \leq p$, and $\lambda_k > 0$ and bounded (for $k = 1, \ldots, p$, and $k \neq i$). Nothing is substantially changed since only the weighted-sum vector direction is important. Both weight normalization procedures are illustrated in Fig. 3.5 for the bi-objective case.

This scalarizing technique can also be applied to integer, mixed-integer and nonlinear programming problems, but it does not allow to obtain unsupported nondominated solutions. Figure 3.6 illustrates the case of integer programming. Solutions B and D are alternative optimal solutions to the weighted-sum objective function with the gradient $f_\lambda$. A slight increase of the weight assigned to $f_1(\mathbf{x})$ leads

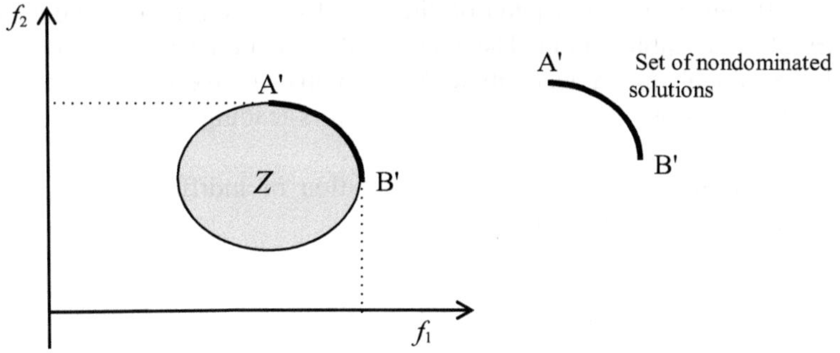


**Fig. 3.7** Nonlinear convex problem

to solution D only and a slight increase of the weight assigned to $f_2(\mathbf{x})$ leads to solution B only; there is no weight vector allowing to reach nondominated solution C, which is unsupported.

In the example of Fig. 3.7, the feasible region Z in the space of the objective functions is convex.

Solutions A′ and B′ are nondominated and it is not possible to obtain them by optimizing $\max_{\mathbf{x}\in X} \{\lambda_1 f_1(\mathbf{x}) + \lambda_2 f_2(\mathbf{x})\}$, with both weights strictly positive. In this problem the solutions A′ and B′ are improper, since $\frac{\partial f_2}{\partial f_1}(\mathrm{A}') = 0$ and $\frac{\partial f_2}{\partial f_1}(\mathrm{B}') = -\infty$, that is, the variation rate of $f_2(\mathbf{x})$ regarding $f_1(\mathbf{x})$ is zero and infinite, respectively (the concept of proper/improper nondominated solution is presented in Chap. 2).

In a (nonlinear) problem in which the feasible region is convex and the objective functions are concave it is possible to compute all proper nondominated solutions using strictly positive weights. If the problem has improper nondominated solutions then these can also be obtained by optimizing a weighted-sum allowing weights equal to zero.

Therefore, since all nondominated solutions in MOLP are proper and supported this scalarizing technique can provide the basis for methods to find the entire set of nondominated solutions (the so-called generating methods).

### *3.2.1 Indifference Regions on the Weight Space in MOLP*

The graphical representation of the weight set that leads to the same basic feasible solution (note that each vertex may correspond to more than one basic solution if degeneracy occurs), is called *indifference region* and can be obtained through the decomposition of the parametric (weight) space $\boldsymbol{\lambda} \in \Lambda = \{\boldsymbol{\lambda} \in \mathbb{R}^p\colon \lambda_k > 0, k = 1, \ldots, p, \sum_{k=1}^{p} \lambda_k = 1\}$. The DM can be "indifferent" to all combinations of weights within this region because they lead to the same efficient solution.

The indifference regions depend on the objective function coefficients and the geometry of the feasible region. The analysis of the parametric (weight) space can be used as a valuable tool to learn about the geometry of the efficient/nondominated region in MOLP, since it gives the weight vectors leading to each efficient basic solution.

Let us start by exemplifying the computation of indifference regions for a problem with two objective functions:

$$\begin{aligned} \max z_1 &= f_1(\mathbf{x}) = 5x_1 + 3x_2 \\ \max z_2 &= f_2(\mathbf{x}) = 2x_1 + 8x_2 \end{aligned}$$

s. t.

$$\left.\begin{aligned} x_1 + 4x_2 &\leq 100 \\ 3x_1 + 2x_2 &\leq 150 \\ 5x_1 + 3x_2 &\geq 200 \\ x_1, x_2 &\geq 0 \end{aligned}\right\} \text{(feasible region } X\text{)}$$

In Fig. 3.8, [AC] represents the set of efficient solutions and [A′C′] represents the corresponding set of nondominated solutions. The slope of [A′C′] is −20. The slope of the level lines of the weighted-sum objective functions, $\lambda_1 f_1(\mathbf{x}) + \lambda_2 f_2(\mathbf{x})$, in the objective function space, is given by $-\frac{\lambda_1}{\lambda_2}$. We consider the weights are normalized: $\sum_{k=1}^{2} \lambda_k = 1$, i.e., $\lambda_2 = 1 - \lambda_1$. Then, the indifference regions associated

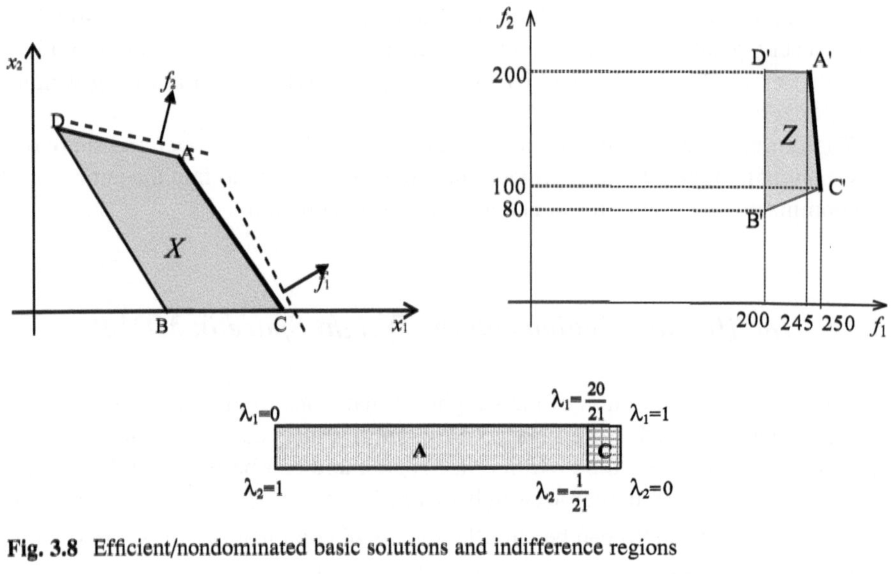


Fig. 3.8 Efficient/nondominated basic solutions and indifference regions

with vertex A and vertex C, i.e., the sets of weights for which solving the problem $\max_{\mathbf{x}\in X}\{\lambda_1 f_1(\mathbf{x}) + \lambda_2 f_2(\mathbf{x})\}$ leads to A and C, respectively, are obtained with $\lambda_1 \in \left(0, \frac{20}{21}\right]$ and $\lambda_2 \in \left[\frac{1}{21}, 1\right)$ for vertex A, and $\lambda_1 \in \left[\frac{20}{21}, 1\right)$ and $\lambda_2 \in \left(0, \frac{1}{21}\right]$ for vertex C.

For the weight values $\lambda_1 = \frac{20}{21}$ and $\lambda_2 = \frac{1}{21}$, points A and C are obtained simultaneously, since $\max_{\mathbf{x}\in X}\{\lambda_1 f_1(\mathbf{x}) + \lambda_2 f_2(\mathbf{x})\}$ leads to edge [AC] as alternative optima.

The determination of the indifference regions in the parametric (weight) diagram for this bi-objective problem was carried out just using the information derived from the geometry of the problem. However, the computation of indifference regions can be done using the multiobjective simplex tableau (i.e., with one reduced cost row for each objective function). In particular, the study of problems with three-objective functions allows a meaningful graphical representation of indifference regions using the information available in the simplex tableau corresponding to a basic (vertex) solution as a result of optimizing a weighted-sum scalarizing function $\max_{\mathbf{x}\in X}\{\lambda_1 f_1(\mathbf{x}) + \lambda_2 f_2(\mathbf{x}) + \lambda_3 f_3(\mathbf{x})\}$ with a given weight vector.

The simplex tableau associated with an efficient basic solution offers the information needed to compute the locus of the weights $\lambda_k$ ($k = 1, \ldots, p$) for which the solution to the weighted-sum problem

$$\begin{aligned} &\max \sum_{k=1}^{p} \lambda_k f_k(\mathbf{x}) \\ &\text{s.t. } \mathbf{x} \in X \end{aligned}$$

with $\boldsymbol{\lambda} \in \Lambda = \{\boldsymbol{\lambda} \in \mathbb{R}^p: \lambda_k > 0,\ k = 1, \ldots, p,\ \sum_{k=1}^{p} \lambda_k = 1\}$ leads to the same efficient basic solution.

For a single objective LP a basic feasible solution to

$$\begin{aligned} &\max z = \mathbf{cx} \\ &\text{s. t. } \mathbf{x} \in X = \{\mathbf{x} \in \mathbb{R}^n : \mathbf{Ax} = \mathbf{b}, \mathbf{x} \geq \mathbf{0}\} \end{aligned}$$

is optimal if and only if $\mathbf{uA} - \mathbf{c} \geq \mathbf{0}$, where the elements of the vector $(\mathbf{uA} - \mathbf{c})$ are called reduced costs (in general, the last row of the simplex tableau). $\mathbf{u} = \mathbf{c_B B}^{-1}$ is a row vector (of dimension $m$), whose elements are the dual variables, $\mathbf{B}$ is the basis matrix corresponding to the current tableau (a sub-matrix $m \times m$ of A, with rank $m$) and $\mathbf{c_B}$ is a sub-vector of $\mathbf{c}$, with dimension $1 \times m$, corresponding to the basic variables.

In MOLP the multiobjective simplex tableau includes a reduced cost row associated with each objective function:

| | |
|---|---|
| $\mathbf{A}$ | $\mathbf{b}$ |
| $-\mathbf{C}$ | $\mathbf{0}$ |

With respect to basis **B** this tableau can be transformed as:

| $\mathbf{x_N}$ | $\mathbf{x_B}$ | |
|---|---|---|
| $\mathbf{B^{-1}N}$ | $\mathbf{I}$ | $\mathbf{B^{-1}\,b}$ |
| $\mathbf{C_B B^{-1}N - C_N}$ | $\mathbf{0}$ | $\mathbf{C_B B^{-1} b}$ |

where **B** and **N**, $\mathbf{C_B}$ and $\mathbf{C_N}$ are the sub-matrices of **A** and **C** corresponding to the basic ($\mathbf{x_B}$) and nonbasic ($\mathbf{x_N}$) variables, respectively. $\mathbf{W} = \mathbf{C_B B^{-1} N} - \mathbf{C_N}$ is the reduced cost matrix associated with basis **B**.

In MOLP the set of weights $\boldsymbol{\lambda}$ for which the basic solution associated with the multiobjective simplex tableau is optimal to the weighted-sum problem is then given by $\{\boldsymbol{\lambda}\mathbf{W} \geq \mathbf{0}, \boldsymbol{\lambda} \in \Lambda\}$.

**Definition of efficient basis**

**B** is an efficient basis if and only if it is an optimal basis to the weighted-sum problem (3.2) for some weight vector $\boldsymbol{\lambda} \in \Lambda$, that is, **B** is an efficient basis if and only if the system $\{\boldsymbol{\lambda}\mathbf{W} \geq \mathbf{0}, \boldsymbol{\lambda} \in \Lambda\}$ is consistent.

**Definition of efficient nonbasic variable**

The nonbasic variable $x_j$ is efficient with respect to basis **B** if and only if $\boldsymbol{\lambda} \in \Lambda$ exists such that

$$\begin{aligned} \boldsymbol{\lambda}\mathbf{W} &\geq \mathbf{0} \\ \boldsymbol{\lambda}\mathbf{W}_{\cdot j} &= 0 \end{aligned}$$

where $\mathbf{W}_{\cdot j}$ is the column vector of **W** corresponding to $x_j$ (that is, the reduced cost of the weighted-sum function associated with $x_j$ can be zero).

The definition of efficient nonbasic variable means that, for a given efficient basis, if $x_j$ is an efficient nonbasic variable then any feasible pivot operation associated with $x_j$ as entering variable leads to an adjacent efficient basis (i.e., obtained from the previous basis through the pivot operation). If the pivot operation leading from one basis $\mathbf{B}_1$ to an adjacent basis $\mathbf{B}_2$ is non-degenerate then the vertices of the feasible region associated with those bases are different and the edge that connects them is composed by efficient solutions. As the bases (vertices) are connected, then it is possible to develop a multiobjective simplex method as an extension of the (single objective) simplex method (Steuer 1986), using sub-problems to test the efficiency of nonbasic variables. This multiobjective simplex method is aimed at computing all efficient bases (vertices). Using this information it is also able to characterize efficient edges and efficient faces (of different dimensions).

The element $w_{kj}$ of the reduced cost matrix **W** represents the rate of change of objective function $f_k(\mathbf{x})$ due to a unit change of the nonbasic variable $x_j$ that becomes basic. Each column of **W** associated with an efficient nonbasic variable represents the rate of change of the objective functions along the corresponding efficient edge emanating from the current vertex.

From the multiobjective simplex tableau corresponding to an efficient basic solution to the MOLP problem, the set of corresponding weights is defined by $\{\boldsymbol{\lambda}\mathbf{W} \geq \mathbf{0},\ \boldsymbol{\lambda} \in \Lambda\}$ thus defining the indifference region. A common frontier to two indifference regions means that the corresponding efficient basic solutions are connected by an efficient edge, which is associated with an efficient nonbasic variable becoming a basic variable. If a point $\boldsymbol{\lambda} \in \Lambda$ belongs to several indifference regions, this means that these regions are associated with efficient solutions located on the same efficient face (this face is only weakly efficient if that point is located on the frontier of the parametric diagram, i.e., some weight $\lambda_\kappa = 0$, $k = 1, \ldots, p$).

The decomposition of the parametric (weight) diagram into indifference regions, i.e., the graphical representation of the set of weights $\boldsymbol{\lambda}$ leading to the same efficient basic solution is especially interesting in problems with three objective functions. Note that due to the normalization condition $\lambda_1 + \lambda_2 + \ldots + \lambda_p = 1$, $\Lambda$ can be represented in a diagram of dimension $p-1$. For the three objective case, the weight diagram $\Lambda = \{\boldsymbol{\lambda} \in \mathbb{R}^3\colon \lambda_k > 0,\ k = 1, 2, 3, \text{ and } \sum_{k=1}^{3} \lambda_k = 1\}$ can be displayed using the equilateral triangle in Fig. 3.9a. Since $\sum_{k=1}^{3} \lambda_k = 1$ the diagram corresponding to the equilateral triangle, defined by the points $\boldsymbol{\lambda} = (1, 0, 0)$, $\boldsymbol{\lambda} = (0, 1, 0)$ and $\boldsymbol{\lambda} = (0, 0, 1)$, can be projected, for example, onto the plane $(0, \lambda_1, \lambda_2)$, without loss of information (Fig. 3.9b).

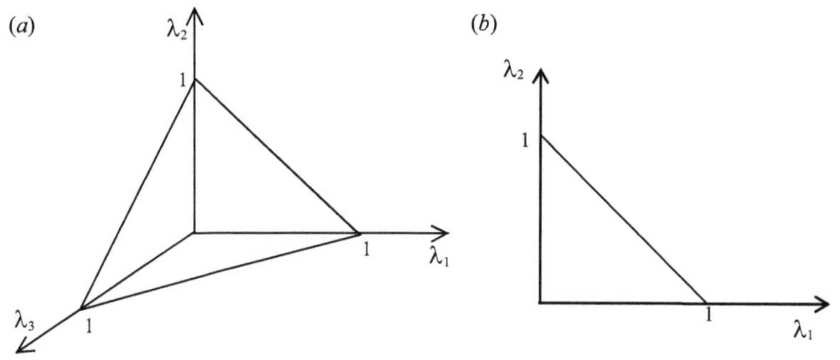


**Fig. 3.9** Parametric (weight) diagram

**Example 1**

$$\max \mathbf{C}\,\mathbf{x} = \begin{bmatrix} 3 & 1 & 2 & 1 \\ 1 & -1 & 2 & 4 \\ -1 & 5 & 1 & 2 \end{bmatrix} \begin{bmatrix} x_1 \\ x_2 \\ x_3 \\ x_4 \end{bmatrix}$$

s. t.

$$X \equiv \begin{cases} 2x_1 + x_2 + 4x_3 + 3x_4 \le 60 \\ 3x_1 + 4x_2 + x_3 + 2x_4 \le 60 \\ x_1 \ge 0, x_2 \ge 0, x_3 \ge 0, x_4 \ge 0 \end{cases}$$

Let us compute the indifference region associated with the efficient basic solution that optimizes the weighted-sum problem with equal weights: $\max\limits_{\mathbf{x}\in X} \left\{ \frac{1}{3} f_1(\mathbf{x}) + \frac{1}{3} f_2(\mathbf{x}) + \frac{1}{3} f_3(\mathbf{x}) \right\}$:

$$\max_{\mathbf{x}\in X} \left\{ \frac{1}{3}(3x_1 + x_2 + 2x_3 + x_4) + \frac{1}{3}(x_1 - x_2 + 2x_3 + 4x_4) + \frac{1}{3}(-x_1 + 5x_2 + x_3 + 2x_4) \right\}$$

$$\max_{\mathbf{x}\in X} \left\{ x_1 + \frac{5}{3}x_2 + \frac{5}{3}x_3 + \frac{7}{3}x_4 \right\}$$

To build the multiobjective simplex tableau a reduced cost row for each objective function is added.

The problem $\max\limits_{\mathbf{x}\in X} \left\{ \frac{1}{3} f_1(\mathbf{x}) + \frac{1}{3} f_2(\mathbf{x}) + \frac{1}{3} f_3(\mathbf{x}) \right\}$ is solved and the reduced cost rows corresponding to the objective functions of the original problem are updated in each pivot operation. $s_1$ and $s_2$ denote the slack variables associated with the constraints.

$z_j^\lambda - c_j^\lambda$ denotes the reduced cost row of the weighted-sum objective function while $z_j^k - c_j^k$ denotes the reduced cost row of each objective function $f_k(\mathbf{x})$. Note that $z_j^\lambda - c_j^\lambda$ is the weighted-sum of $z_j^k - c_j^k$ ($k = 1, 2, 3$), i.e., $z_j^\lambda - c_j^\lambda = \sum_{k=1}^{3} \lambda_k \left( z_j^k - c_j^k \right)$ for all nonbasic variables $x_j$.

The initial tableau is:

| | | $x_1$ | $x_2$ | $x_3$ | $x_4$ | $s_1$ | $s_2$ | |
|---|---|---|---|---|---|---|---|---|
| | $s_1$ | 2 | 1 | 4 | 3 | 1 | 0 | 60 |
| | $s_2$ | 3 | 4 | 1 | 2 | 0 | 1 | 60 |
| Reduced cost row of the weighted–sum objective function | $z_j^\lambda - c_j^\lambda$ | $-1$ | $-\frac{5}{3}$ | $-\frac{5}{3}$ | $-\frac{7}{3}$ | 0 | 0 | 0 |
| Reduced cost matrix | $z_1 - c_1$ | $-3$ | $-1$ | $-2$ | $-1$ | 0 | 0 | 0 |
| | $z_2 - c_2$ | $-1$ | 1 | $-2$ | $-4$ | 0 | 0 | 0 |
| | $z_3 - c_3$ | 1 | $-5$ | $-1$ | $-2$ | 0 | 0 | 0 |

The optimal tableau associated with the basic solution that solves $\max_{\mathbf{x}\in X}\left\{\frac{1}{3}f_1(\mathbf{x})+\frac{1}{3}f_2(\mathbf{x})+\frac{1}{3}f_3(\mathbf{x})\right\}$ is

| | $x_1$ | $x_2$ | $x_3$ | $x_4$ | $s_1$ | $s_2$ | |
|---|---|---|---|---|---|---|---|
| $x_4$ | $\frac{1}{2}$ | 0 | $\frac{3}{2}$ | 1 | $\frac{2}{5}$ | $-\frac{1}{10}$ | 18 |
| $x_2$ | $\frac{1}{2}$ | 1 | $-\frac{1}{2}$ | 0 | $-\frac{1}{5}$ | $\frac{3}{10}$ | 6 |
| $z_j^\lambda - c_j^\lambda$ | 1 | 0 | 1 | 0 | $\frac{3}{5}$ | $\frac{4}{15}$ | 52 |
| $z_1 - c_1$ | $-2$ | 0 | $-1$ | 0 | $\frac{1}{5}$ | $-\frac{1}{5}$ | 24 |
| $z_2 - c_2$ | $\frac{1}{2}$ | 0 | $\frac{9}{2}$ | 0 | $\frac{9}{5}$ | $\frac{7}{10}$ | 66 |
| $z_3 - c_3$ | $\frac{9}{2}$ | 0 | $-\frac{1}{2}$ | 0 | $-\frac{1}{5}$ | $\frac{13}{10}$ | 66 |

The set of weights $(\lambda_1, \lambda_2, \lambda_3) > 0$, with $\lambda_3 = 1 - \lambda_1 - \lambda_2$, for which this basic solution is optimal to the weighted-sum scalar problem and therefore efficient to the MOLP problem is given by

$$\begin{cases} -2\lambda_1 + \frac{1}{2}\lambda_2 + \frac{9}{2}\,(1-\lambda_1-\lambda_2) \geq 0 & \text{(a)} \\ -\lambda_1 + \frac{9}{2}\lambda_2 - \frac{1}{2}\,(1-\lambda_1-\lambda_2) \geq 0 & \text{(b)} \\ \frac{1}{5}\lambda_1 + \frac{9}{5}\lambda_2 - \frac{1}{5}\,(1-\lambda_1-\lambda_2) \geq 0 & \text{(c)} \\ \frac{1}{5}\lambda_1 - \frac{7}{10}\lambda_2 + \frac{13}{10}\,(1-\lambda_1-\lambda_2) \geq 0 & \text{(d)} \end{cases}$$

Each of these constraints in $(\lambda_1, \lambda_2, \lambda_3)$ is associated with a nonbasic variable w. r. t. the present basis. This set of constraints can be written as a function of $(\lambda_1, \lambda_2)$:

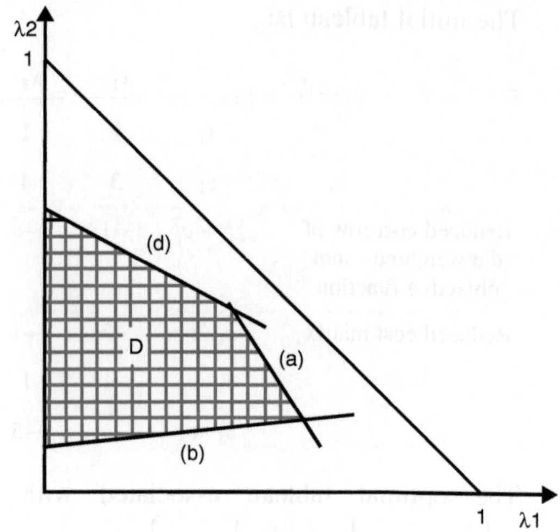


**Fig. 3.10** Indifference region associated with efficient basic solution D

$$\begin{cases} \dfrac{13}{2}\lambda_1 + 4\lambda_2 \leq \dfrac{9}{2} & \text{(a)} \\ -\dfrac{1}{2}\lambda_1 + 5\lambda_2 \geq \dfrac{1}{2} & \text{(b)} \\ \dfrac{2}{5}\lambda_1 + 2\lambda_2 \geq \dfrac{1}{5} & \text{(c)} \\ \dfrac{11}{10}\lambda_1 + 2\lambda_2 \leq \dfrac{13}{10} & \text{(d)} \end{cases}$$

Constraints (a), (b) and (d) delimit the indifference region D (Fig. 3.10). Therefore, the corresponding nonbasic variables $x_1$, $x_3$ and $s_2$ are efficient. Any weight vector $\boldsymbol{\lambda}\in\Lambda$ satisfying these constraints leads to a weighted-sum problem $\max_{\mathbf{x}\in X} \{\lambda_1 f_1(\mathbf{x}) + \lambda_2 f_2(\mathbf{x}) + \lambda_3 f_3(\mathbf{x})\}$ whose optimal solution is the same efficient basis, i.e., the one determined by optimizing the problem $\max_{\mathbf{x}\in X} \left\{\frac{1}{3}f_1(\mathbf{x}) + \frac{1}{3}f_2(\mathbf{x}) + \frac{1}{3}f_3(\mathbf{x})\right\}$. Constraint (c) is redundant, i.e., it does not contribute to define the indifference region associated with the efficient solution $(x_4, x_2) = (18, 6)$, $(z_1, z_2, z_3) = (24, 66, 66)$. Therefore, the corresponding nonbasic variable $s_1$ is not efficient.

Constraints (a), (b) and (d) correspond to edges of the efficient region emanating from the current vertex, solution D. "Crossing" these constraints leads to efficient vertices adjacent to D. For instance, making $x_1$ a basic variable leads to an adjacent efficient basic solution (vertex) whose indifference region is contiguous to D and delimited by constraint $\frac{13}{2}\lambda_1 + 4\lambda_2 \geq \frac{9}{2}$ (opposite to (a) above). Therefore, $x_1$, as well as $x_3$ and $s_2$ (but not $s_1$), are efficient nonbasic variables because, when becoming basic variables, each one leads to an adjacent efficient vertex solution (with the corresponding indifference region) through an efficient edge, i.e., an edge composed of efficient solutions only. Note that there may exist an edge connecting two efficient vertex solutions that is not composed of efficient solutions.

Efficient nonbasic variables can be identified in the multiobjective simplex tableau. In problems with two or three objective functions, efficient nonbasic variables can also be recognized from the display of indifference regions in the parametric (weight) diagram.

## 3.3 Minimizing a Distance/Achievement Function to a Reference Point

The minimization of the distance, according to a certain metric, of the feasible region to a reference point defined in the objective function space can be used to compute nondominated solutions.

The ideal solution $\mathbf{z}^*$ is often used as reference point. The rationale is that it offers the best value for each evaluation dimension reachable in the feasible region, since its components result from optimizing individually each objective function. If the reference point represents an attainable outcome, the scalarizing function is called an *achievement* scalarizing function, as it aims to reach or surpass the reference point.

### *3.3.1 A Brief Review of Metrics*

A metric is a distance function that assigns a scalar $\|\mathbf{z}^1 - \mathbf{z}^2\| \in \mathbb{R}$ to each pair of points $\mathbf{z}^1, \mathbf{z}^2 \in \mathbb{R}^n$ (where n is the dimension of the space).

For the $L_q$ metric the distance between two points in $\mathbb{R}^n$ is given by:

$$\|\mathbf{z}^1 - \mathbf{z}^2\|_q = \left[\sum_{i=1}^{n} |z_i^1 - z_i^2|^q\right]^{1/q} \qquad q \in \{1, 2, \ldots\}$$

$$\|\mathbf{z}^1 - \mathbf{z}^2\|_\infty = \max_{i=1,\ldots,n} |z_i^1 - z_i^2|$$

The loci of the points at the same distance from $\mathbf{z}^*$ (isodistance contour), according to the metrics $L_1$, $L_2$ and $L_\infty$ are displayed in Fig. 3.11.

In Fig. 3.12 a, b and c, points A, B and C minimize the distance of region Z to $\mathbf{z}^*$, using the metrics $L_1$, $L_2$ and $L_\infty$, respectively.

The metrics $L_1$, $L_2$ and $L_\infty$ are especially important. $L_1$ is the sum of all components of $|\mathbf{z}^1 - \mathbf{z}^2|$, i.e., the *city block* distance in a "rectangular city" as Manhattan. $L_2$ is the Euclidean distance. $L_\infty$ is the Chebyshev distance in which

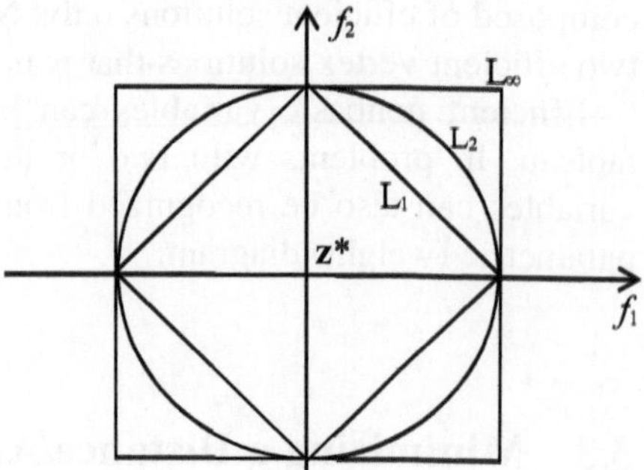


**Fig. 3.11** Loci of equidistant points from z* for $L_1$, $L_2$ and $L_\infty$ metrics

only the worst case is considered, i.e., the largest difference component in $|\mathbf{z}^1 - \mathbf{z}^2|$.

A weighted family of $L_q^\lambda$ metrics can also be defined, where the vector $\boldsymbol{\lambda} \geq \mathbf{0}$ is used to assign a different scale (or "importance") factor to the multiple components:

$$\left\|\mathbf{z}^1 - \mathbf{z}^2\right\|_q^\lambda = \left[\sum_{i=1}^{n} \left(\lambda_i \left|z_i^1 - z_i^2\right|\right)^q\right]^{1/q} \qquad q \in \{1, 2, \ldots\}$$

$$\left\|\mathbf{z}^1 - \mathbf{z}^2\right\|_\infty^\lambda = \max_{i=1,\ldots,n} \lambda_i \left|z_i^1 - z_i^2\right|$$

The loci of points at the same distance of **z***, according to the weighted $L_1^\lambda$, $L_2^\lambda$, and $L_\infty^\lambda$ metrics are illustrated in Fig. 3.13, representing the isodistance contour for each metric with $\lambda_1 < \lambda_2$.

The external isodistance contour $L_\infty^{\lambda,\rho}$ presented in Fig. 3.14 regards to $\left\|\mathbf{z}^1 - \mathbf{z}^2\right\|_\infty^\lambda + \sum_{i=1}^{2} \rho_i \left|z_i^1 - z_i^2\right|$, with a small positive $\rho_i$, which can be seen as a combination of $L_\infty^\lambda$ and $L_1^\lambda$ metrics. This is generally called the augmented weighted Chebyshev metric, $L_\infty^{\lambda,\rho}$.

Although $L_q^\lambda$, for $q \in \{1, 2, \ldots\}$, can be used to determine nondominated solutions by solving scalar problems involving the minimization of a distance to a reference point, we will formally present only the case using the Chebyshev metric ($L_\infty^\lambda$). This metric is important since it captures the attitude of minimizing the largest difference, i.e., the worst deviation, to the value that is desired in all evaluation dimensions. In general, the $L_\infty^{\lambda,\rho}$ metric is used to guarantee that the solutions obtained are nondominated and not just weakly nondominated.

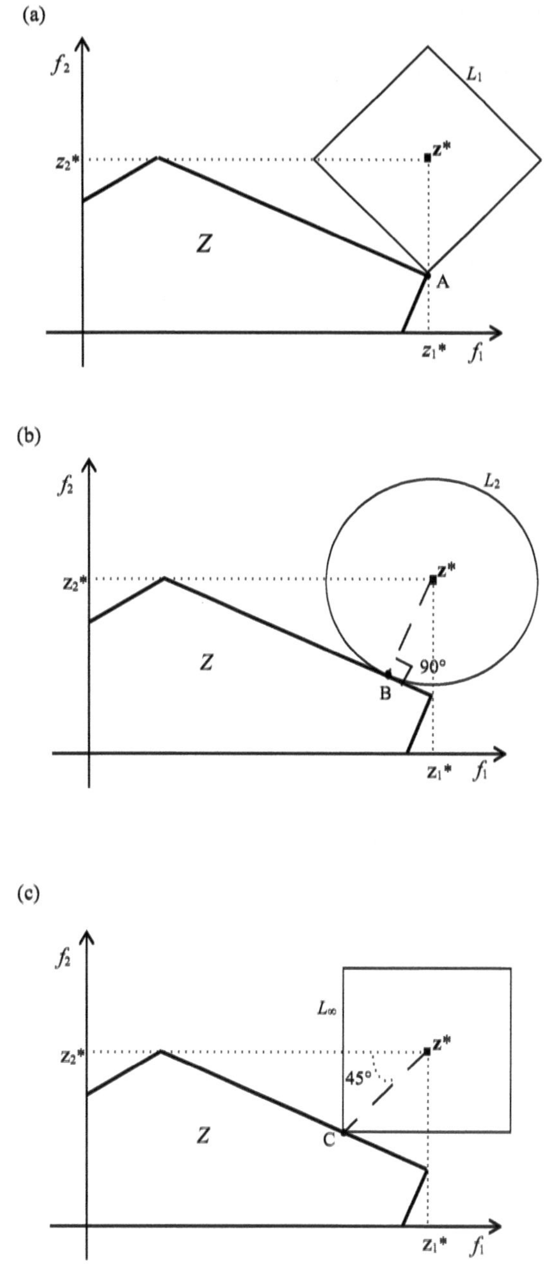


**Fig. 3.12** Nondominated solutions that minimize the distance to the ideal solution according to $L_1$, $L_2$ and $L_\infty$ metrics. (**a**) Point A minimizes the distance to z* according to the $L_1$ metric. (**b**) Point B minimizes the distance to z* according to the $L_2$ metric. (**c**) Point C minimizes the distance to z* according to the $L_\infty$ metric

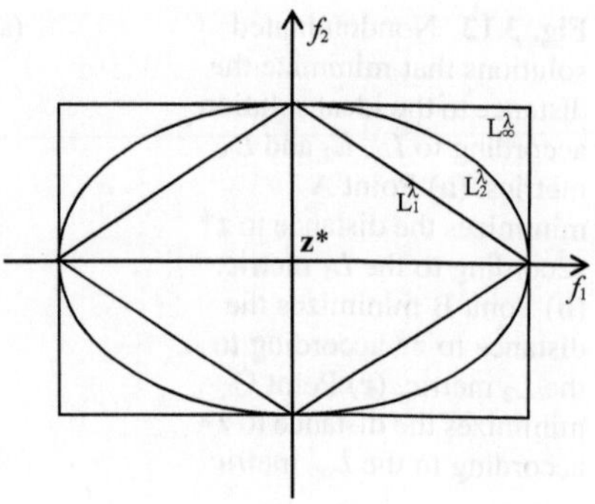


**Fig. 3.13** Loci of points equidistant from z*, for $L_1^\lambda$, $L_2^\lambda$ and $L_\infty^\lambda$ metrics

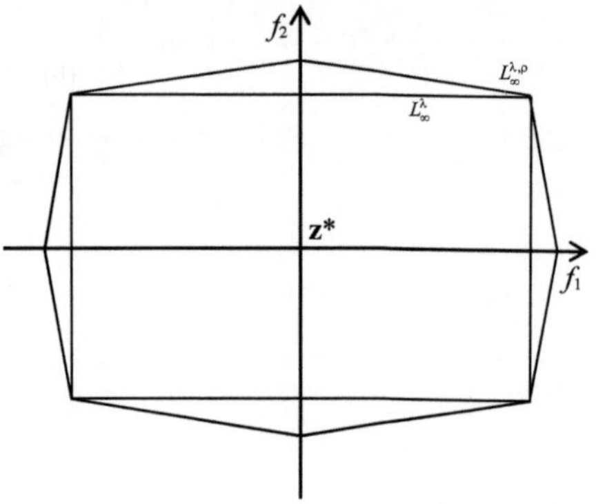


**Fig. 3.14** Loci of points equidistant from z*, for $L_\infty^\lambda$ and $L_\infty^{\lambda,\rho}$

**Proposition 3**
If $\mathbf{x}^1 \in X$ is a solution to the problem (3.3)

$$\min_{\mathbf{x}\in X} \left\{ \max_{k=1,\ldots,p} \lambda_k \left[ z_k^* - f_k(\mathbf{x}) \right] - \sum_{k=1}^{p} \rho_k f_k(\mathbf{x}) \right\}, \text{ for some } \boldsymbol{\lambda} \geq \mathbf{0} \qquad (3.3)$$

where the $\rho_k$ are small positive scalars, then $\mathbf{x}^1$ is an efficient solution to the multiobjective problem.

This surrogate problem entails determining the feasible solution that minimizes the distance based on an augmented weighted Chebyshev metric to the ideal solution.

The truthfulness of Proposition 3 can be shown as follows. Let us suppose that $\mathbf{x}^1$ is not an efficient solution and is optimal to problem (3.3), with $\max_{k=1,\ldots,p} \lambda_k \left[ z_k^* - f_k(\mathbf{x}) \right] = v_1$. Therefore, there is a solution $\mathbf{x}^2$ such that $f_k(\mathbf{x}^2) \geq f_k(\mathbf{x}^1)$, $k = 1,\ldots, p$, and the inequality is strict for at least one $k$. In these circumstances, $\lambda_k \left( z_k^* - f_k(\mathbf{x}^2) \right) \leq v_1$, $k = 1, \ldots, p$, and

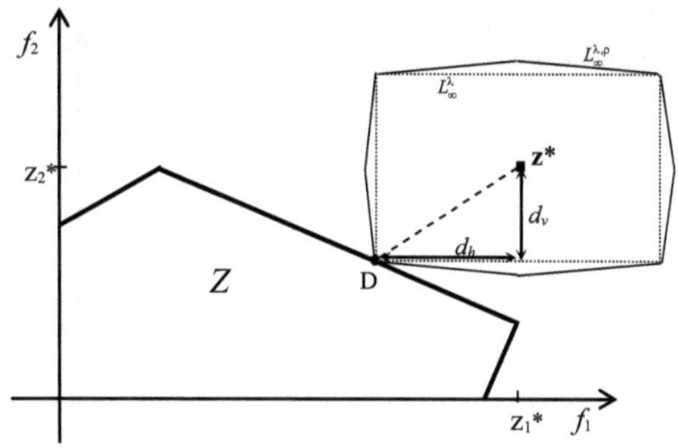


**Fig. 3.15** Minimizing the augmented weighted Chebyshev distance

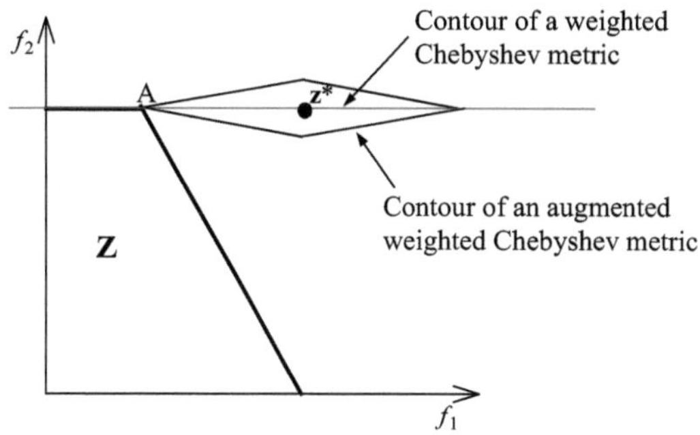


**Fig. 3.16** Illustration of the weighted Chebyshev metric and the augmented weighted Chebyshev metric

$\sum_{k=1}^{p} \rho_k f_k(\mathbf{x}^2) > \sum_{k=1}^{p} \rho_k f_k(\mathbf{x}^1)$. Hence, $\mathbf{x}^1$ would not be optimal to (3.3), which contradicts the hypothesis. Therefore, $\mathbf{x}^1$ has to be efficient.

Proposition 3 also establishes a necessary condition for efficiency in MOLP, for sufficiently small $\rho_k$. Therefore, this scalarizing technique enables to obtain the entire set of efficient/nondominated solutions to a MOLP problem.

The computation process is illustrated in Fig. 3.15. Point D is the solution that minimizes the distance to $\mathbf{z}^*$according to $L_\infty^{\lambda,\rho}$ (3.3) or $L_\infty^{\lambda}$ ((3.3) without the term $\sum_{k=1}^{p} \rho_k f_k(\mathbf{x})$), considering a particular weight vector $\boldsymbol{\lambda}$, where $\lambda_1 < \lambda_2$. Note that $\frac{d_v}{d_h} = \frac{\lambda_1}{\lambda_2}$.

The term $\sum_{k=1}^{p} \rho_k f_k(\mathbf{x})$ is used to avoid solutions that are only weakly efficient when the scalarizing problem (3.3) has alternative optimal solutions (Fig. 3.16). Considering $\lambda_1 = 0$, all solutions on the horizontal line passing through $\mathbf{z}^*$ are equidistant from $\mathbf{z}^*$ according to $L_\infty^{\lambda}$. The consideration of $L_\infty^{\lambda,\rho}$ enables to obtain the strictly nondominated solution A.

Problem (3.3) is equivalent to the programming problem (3.4):

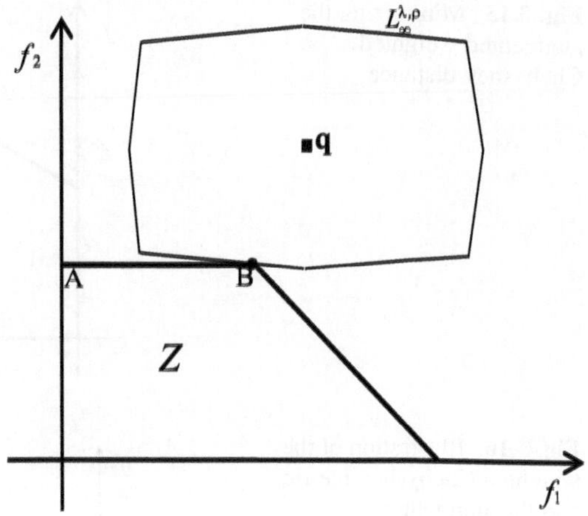


**Fig. 3.17** Minimizing an augmented weighted Chebyshev distance to a reference point

$$\min \left\{ v - \sum_{k=1}^{p} \rho_k f_k(\mathbf{x}) \right\}$$

s.t.

$$\begin{aligned} &\lambda_k\left(z_k^* - f(\mathbf{x})\right) \leq v \quad k = 1, \ldots, p \\ &\mathbf{x} \in X \\ &v \geq 0 \end{aligned} \tag{3.4}$$

Other reference points can be used. The components of the reference point represent the values that the DM would like to attain for each objective function. For this purpose the ideal solution **z*** is interesting because its components are the best values that can be reached for each objective function in the feasible region.

Figure 3.17 displays the computation of nondominated solutions by minimizing an augmented weighted Chebyshev distance to the unattainable reference point **q**. In the example of Fig. 3.17, minimizing a (non-augmented) Chebyshev distance would enable to compute all solutions on segment [AB]; however, these solutions are just weakly nondominated except solution B, which is strictly nondominated.

If the reference point is in the interior of the feasible region, i.e., it is dominated, it is no longer possible to consider a Chebyshev metric, but this technique for computing nondominated solutions is still valid considering the auxiliary variable $v$ unrestricted in sign in problem (3.4). In these circumstances, this is an *achievement* scalarizing program. Figure 3.18 shows the projection of a reference point **q** located in the interior of the feasible region onto the nondominated frontier. $\mathbf{z}^1$ is the nondominated solution that is obtained through the resolution of problem (3.4), with **q** replacing **z*** and considering variable $v$ unrestricted in sign. In the case illustrated in Fig. 3.18, $\lambda_1 < \lambda_2$.

This technique for computing nondominated solutions is valid for more general cases than MOLP.

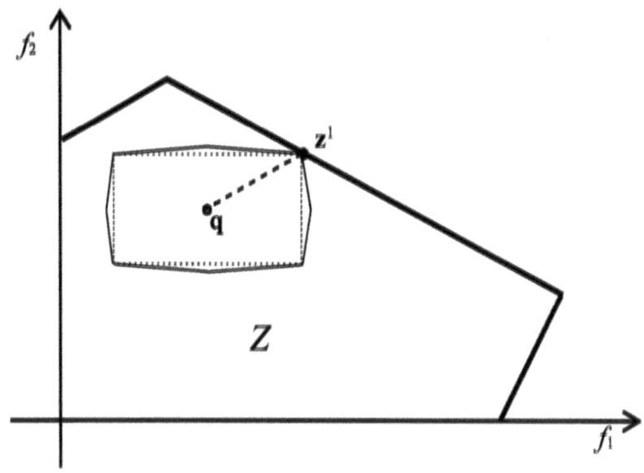


**Fig. 3.18** Projection of a dominated reference point located in the interior of the feasible region onto the nondominated frontier

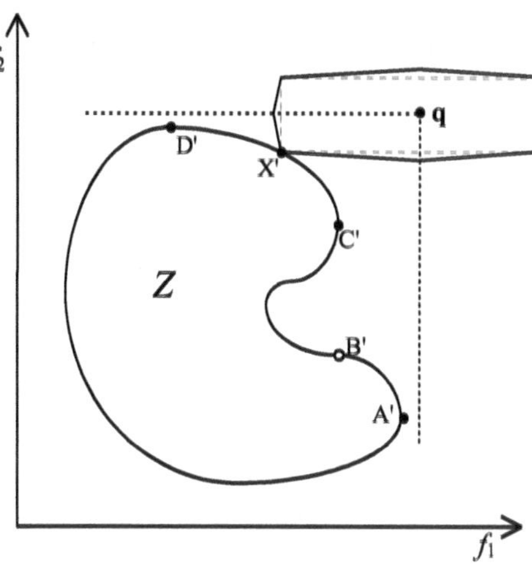


**Fig. 3.19** Finding nondominated solutions in a nonlinear problem using the augmented weighted Chebyshev metric

Figure 3.19 illustrates a nonlinear problem: the point that minimizes the distance to the reference point **q**, using the augmented weighted Chebyshev metric, is a nondominated solution (point X′). Changing the weights, all nondominated solutions can be computed, i.e., solutions on the arcs B′A′, except B′ (which is weakly nondominated) and C′D′. Note that $\mathbf{q} > \mathbf{z}^{*}$ is necessary to guarantee that D′ and A′ are obtained.

Figure 3.20 illustrates the integer programming case. Using the (augmented) weighted Chebyshev metric it is possible to reach all nondominated solutions including unsupported solution C. Note that C was unreachable by optimizing weighted sums of the objective functions (Fig. 3.6).

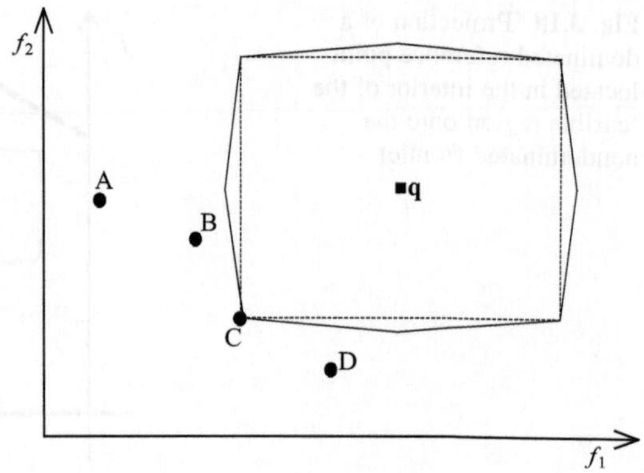


**Fig. 3.20** The augmented Chebyshev metric in integer programming

## 3.4 Classification of Methods to Compute Nondominated Solutions

Different classifications of MOP methods have been proposed according to several parameters, such as the degree of intervention of the DM, the type of modeling of the DM's preferences, the number of decision makers, the inputs required and the outputs generated, etc. (Cohon 1978; Hwang and Masud 1979; Chankong and Haimes 1983; Steuer 1986).

The classification based on the degree of intervention of the DM establishes, in general, the following categories:

(a) *A priori* articulation of the DM's preferences. Once the method is chosen (and possibly some parameters are fixed), the preference aggregation is established.
(b) Progressive articulation of the DM's preferences. This occurs in interactive methods, which comprise a sequence of computation phases (of nondominated solutions) and dialogue phases. The DM's input in the dialogue phase in face of one solution (or a set of solutions in some cases) generated in the computation phase is used to prepare the next computation phase with the aim to obtain another nondominated solution more in accordance with his/her expressed preferences. The characteristics of the dialogue and computation phases, as well as the stopping conditions of the interactive process, depend on the method. Chap. 4 is entirely devoted to interactive methods.
(c) *A posteriori* articulation of the DM's preferences. This deals with methods for characterizing the entire set of nondominated solutions, and the aggregation of the DM's preferences is made in face of the nondominated solutions obtained.

The classification based on the modeling of the DM's preferences generally considers the establishment of a global utility function, priorities between the objective functions, aspiration levels or targets for the objective functions, pairwise comparisons (either of solutions or objective functions) or marginal rates of substitution.

The classification based on the number of decision makers encompasses the situations where a single or several DM are at stake.

The classification based on the type of inputs required and outputs generated consider the type and reliability of data, the participation of the DM in the modeling phase, either the search for the best compromise nondominated solution or a satisfactory solution, selecting, ranking or clustering the solutions.

Other classifications of MOP methods are used according to the fields of application (e.g., systems engineering, project evaluation, etc.).

## 3.5 Methods Based on the Optimization of an Utility Function

In this type of approach, an utility function $U[f_1(\mathbf{x}), f_2(\mathbf{x}), \ldots, f_p(\mathbf{x})]$ is built. If (concave) function U satisfies certain properties, the optimum of $U[\mathbf{f}(\mathbf{x})]$ belongs to the set of nondominated solutions (Steuer 1986). In Fig. 3.21 an illustrative example is presented.

The curves $U[f_1(\mathbf{x}), f_2(\mathbf{x}), \ldots, f_p(\mathbf{x})] = \kappa^i$, with $\kappa^i$ constant, are called indifference curves, and the point belonging to the nondominated solution set which is tangent to one indifference curve is called compromise point (Fig. 3.21).

The utility functions may have the following structure:

$$U[f_1(\mathbf{x}), f_2(\mathbf{x}), \ldots, f_p(\mathbf{x})] = U_1[f_1(\mathbf{x})] + \ldots + U_p[f_p(\mathbf{x})]$$

The weighted-sum of the objective functions may be faced as a particular case of this utility function structure. The "relative importance" of the objective functions may be taken into account through the assignment of a weight vector. If the problem under study is linear and $U[\mathbf{f}(\mathbf{x})]$ is linear then the problem to be solved is also a single objective LP problem.

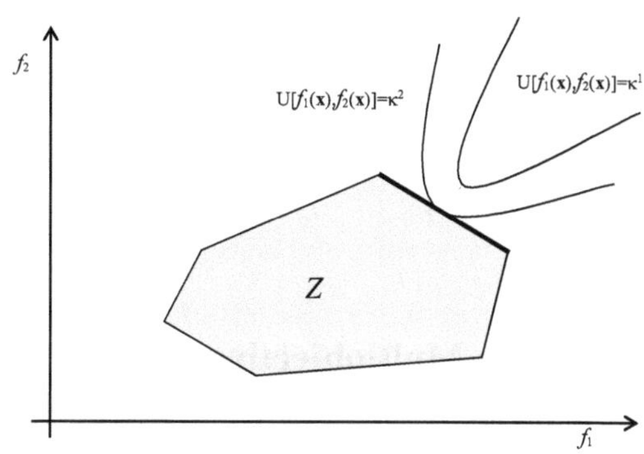


**Fig. 3.21** Example of the use of utility functions

## 3.6 The Lexicographic Method

In this method the objective functions are ranked according to the DM's preferences and then they are sequentially optimized. In each step, an objective function $f_k(\mathbf{x})$ is optimized and an equality constraint is added to the next optimization problems taken into account the optimal value obtained ($f_k(\mathbf{x}) = z_k^*$). In some cases the constraint is not an equality, i.e., deviations w. r. t. the obtained optima are allowed. Note that by adopting the first option, if the optimum of the first ranked objective function is unique, then the procedure stops. This method may be considered an a priori method just requiring ordinal information.

## 3.7 Goal Programming

Goal programming may be viewed as the "bridge" between single objective and MOP, namely concerning reference point approaches. The aim is to minimize a function of the deviations regarding targets ($O_1, \ldots, O_p$) established by the DM for the objective functions. A possible formulation consists in the minimization of a weighted sum of the deviations, with non-negative weights $\alpha_k$ and $\beta_k$:

$$\min \sum_{k=1}^{p} \left(\alpha_k d_k^- + \beta_k d_k^+\right)$$

s. t.

$$\begin{aligned} & f_k(\mathbf{x}) + d_k^- - d_k^+ = O_k && k = 1, \ldots, p \\ & \mathbf{x} \in X \\ & d_k^- \geq 0, d_k^+ \geq 0 && k = 1, \ldots, p \end{aligned}$$

where $d_k^-$ and $d_k^+$ are negative and positive deviations regarding goal $k$, respectively.

The targets established by the DM may lead to a dominated solution to the problem under study if the DM is not sufficiently ambitious in specifying his/her goals. In this case, the goal programming model leads to a satisfactory solution, but it may not belong to the nondominated solution set. For further details about different versions of goal programming see, for example, Steuer (1986) or Romero (1991).

## 3.8 The Multiobjective Simplex Method for MOLP

The algorithms based on the extension of the simplex method for computing the set of efficient vertices (basic solutions) in MOLP can be structured as follows:

(a) Computation of an efficient vertex. For instance, this can be done by optimizing a weighted-sum of the objective functions as explained above.
(b) Computation of the remaining efficient vertices by

1. computing the adjacent efficient bases (Zeleny 1974; Yu and Zeleny 1975; Steuer 1986);
2. computing the adjacent efficient vertices (Evans and Steuer 1973; Steuer 1986).
3. using a parametric technique.

In (b1) and (b2) a theorem presented, for instance, in Yu and Zeleny (1975) is used, which establishes the proof that the set of efficient basic solutions is connected[1]. This means that the entire set of efficient bases (or vertices) can be obtained, by exhaustively examining the adjacent bases of the set of efficient bases that are progressively obtained starting from the initial one, computed in (b).

Steuer (1986) and Zeleny (1974) use an efficiency test to verify whether each basis (or each vertex) under analysis is efficient or not.

Zeleny (1974) establishes several propositions aimed at exploring the maximum information contained in the multiobjective simplex tableau. This is an extension of the simplex method considering one additional row for each objective function and avoiding, whenever possible, unnecessary pivoting operations and the application of the efficiency test.

## 3.9 Proposed Exercises

1. Consider proposed exercise 1, Chap. 2.

   (a) Formulate the problem to determine the solution that minimizes the distance to the ideal solution, according to the $L_\infty$ metric. Obtain graphically and analytically (using an LP solver) the solution to this problem.
   (b) Obtain graphically the solution that minimizes the distance to the ideal solution according to the $L_1$ metric. Identify the efficient nonbasic variables for this solution

2. Consider proposed exercise 2, Chap. 2.

   (a) Find the indifference regions in the parametric (weight) diagram corresponding to the efficient basic solutions that optimize each objective function individually.
   (b) For each solution determined in (a), identify the efficient nonbasic variables

[1] Let $S = \{\mathbf{x}^i : i = 1, \ldots, s\}$ be the set of efficient basic solutions of $X$. This set is connected if it contains only one element or if, for any two points $\mathbf{x}^j, \mathbf{x}^k \in S$, there is a sequence $\{\mathbf{x}^{i_1}, \ldots, \mathbf{x}^{\ell}, \ldots, \mathbf{x}^{i_r}\}$ in $S$, such that $\mathbf{x}^{\ell}$ and $\mathbf{x}^{\ell+1}, \ell = i_1, \ldots, i_{r-1}$, are adjacent and $\mathbf{x}^j = \mathbf{x}^{i_1}, \mathbf{x}^k = \mathbf{x}^{i_r}$.

3. Consider proposed exercise 3, Chap. 2.
   (a) Formulate the problem to determine the efficient solution that minimizes the distance to the ideal solution, according to the $L_l$ metric. Solve this problem graphically.
   (b) Represent qualitatively the parametric (weight) diagram decomposition, considering all the indifference regions.
   (c) Determine graphically the nondominated solution obtained by the *e-constraint* technique when $f_1(\mathbf{x})$ is optimized and $f_2(\mathbf{x}) \geq 3$.

4. Consider the MOLP model with three objective functions:

$$\begin{aligned}
\max \quad & f_1(\mathbf{x}) = 2x_1 + x_2 + 3x_3 + x_4 \\
\max \quad & f_2(\mathbf{x}) = 2x_1 + 4x_2 + x_3 - x_4 \\
\max \quad & f_3(\mathbf{x}) = x_1 + 2x_2 - x_3 + 5x_4 \\
\text{s. t.} \quad & x_1 + 2x_2 + 3x_3 + 4x_4 \leq 40 \\
& 4x_1 + 4x_2 + 2x_3 + x_4 \leq 40 \\
& x_1 \geq 0,\ x_2 \geq 0,\ x_3 \geq 0,\ x_4 \geq 0
\end{aligned}$$

   (a) Find the indifference region corresponding to the efficient solution that optimizes objective function $f_2(\mathbf{x})$.
   (b) What are the efficient nonbasic variables for that solution? Support your analysis on the weight space.
   (c) Consider the following auxiliary problem:

$$\begin{aligned}
\min \quad & \mathrm{v} \\
\text{s. t.} \quad & \mathbf{x} \in X(\text{original feasible region}) \\
& \mathrm{v} + 2x_1 + 4x_2 + x_3 - x_4 \geq 35 \\
& \mathrm{v} + 2x_1 + x_2 + 3x_3 + x_4 \geq 5 \\
& \mathrm{v} + x_1 + 2x_2 - x_3 + 5x_4 \geq 15 \\
& \mathrm{v} \geq 0
\end{aligned}$$

   Is the solution to this auxiliary problem a (strictly) nondominated solution to the multiobjective problem? If not, what changes should be made in the formulation of this auxiliary problem in order to guarantee obtaining a (strictly) nondominated solution?

5. Discuss the following statements, stating whether they are true or false, and presenting a counter example if they are false. Use graphical examples if that facilitates the analysis.

   (a) It is always possible to define $\mathbf{x}^*$ (in the decision space), such that $\mathbf{z}^* = \mathbf{f}(\mathbf{x}^*)$, where $\mathbf{f}(\mathbf{x}) = [f_1(\mathbf{x}), f_2(\mathbf{x}), \ldots, f_p(\mathbf{x})]$.
   (b) In a MOLP problem, the solution obtained by minimizing the distance to a reference point according to the $L_\infty$ metric is always a vertex of the original feasible region.
   (c) In a MOLP problem, the solution obtained by minimizing the distance to a reference point according to the $L_1$ metric is always a vertex of the original feasible region.
   (d) When an additional constraint is introduced into a MOLP problem, it is possible to obtain nondominated solutions to the modified problem that are dominated in the original problem.
   (e) In a MOLP problem with $p$ objective functions it is possible that the whole feasible region is efficient.
   (f) The solutions located on an edge that connects two nondominated vertices are also nondominated.
   (g) Consider a MOLP problem with three objective functions, where three nondominated basic solutions are known. The optimization of a scalarizing function whose gradient is normal to the plane that includes these three solutions always guarantees obtaining a nondominated solution.
   (h) Consider the MOLP problem:

$$\begin{array}{ll} \max & f_1(\mathbf{x}) = \mathbf{c}_1\mathbf{x} \\ \max & f_2(\mathbf{x}) = \mathbf{c}_2\mathbf{x} \\ \max & f_3(\mathbf{x}) = \mathbf{c}_3\mathbf{x} \\ \text{s. t.} & \mathbf{x} \in X \equiv \{\mathbf{A}\mathbf{x} = \mathbf{b}, \mathbf{x} \geq \mathbf{0}\}, \end{array}$$

   where $\mathbf{x}$ $(n \times 1)$, $\mathbf{c}_k$ $(1 \times n)$, $\mathbf{A}$ $(m \times n)$ and $\mathbf{b}$ $(m \times 1)$

   (h.1) Is it possible to obtain a nondominated solution that maximizes $f_3(\mathbf{x})$ with $\lambda_3 = 0$, by solving the weighted-sum problem

$$\begin{array}{ll} \max & \lambda_1\mathbf{c}_1\mathbf{x} + \lambda_2\mathbf{c}_2\mathbf{x} + \lambda_3\mathbf{c}_3\mathbf{x} \\ \text{s. t.} & \mathbf{x} \in X \end{array}$$

   with $(\lambda_1, \lambda_2, \lambda_3) \in \Lambda \equiv \{\lambda_k \geq 0,\ k = 1, 2, 3, \text{ and } \lambda_1 + \lambda_2 + \lambda_3 = 1\}$?

   If this assertion is true, then represent one possible parametric (weight) diagram decomposition in this condition.

(h.2) In what conditions is it possible to guarantee that the optimal solution to

$$\begin{array}{ll} \max & \lambda_1 \mathbf{c}_1\mathbf{x} + \lambda_2 \mathbf{c}_2\mathbf{x} + \lambda_3 \mathbf{c}_3\mathbf{x} \\ \text{s. t.} & \mathbf{x} \in X \end{array}$$

is a nondominated solution to the three objective original problem?

6. Consider the following MOLP problem:

$$\begin{array}{ll} \max f_1(\mathbf{x}) = & 1.5\,x_1 + x_2 \\ \max f_2(\mathbf{x}) = & \qquad\quad x_2 + 2x_3 \\ \max f_3(\mathbf{x}) = & x_1 + x_2 + x_3 \\ \text{s.t.} & x_1 + x_2 + x_3 \leq 8 \\ & x_1 \qquad\;\; + 2x_3 \leq 2 \\ & x_1 \geq 0, x_2 \geq 0, x_3 \geq 0 \end{array}$$

(a) Compute the nondominated solution that maximizes a weighted-sum of the objective functions assigning equal weight to all objectives.
(b) Represent the corresponding indifference region in the parametric (weight) diagram.
(c) Compute a nondominated solution which is adjacent to the one computed in (a) improving objective function $f_1(\mathbf{x})$.
(d) Specify the values of the objective functions of a nondominated nonbasic solution, whose value for $f_1(\mathbf{x})$ is an intermediate value between the ones of the solutions obtained in (a) and in (c).

7. Consider the following MOLP problem:

$$\begin{array}{lll} \max & f_1(\mathbf{x}) = & \qquad x_2 \\ \max & f_2(\mathbf{x}) = & x_1 + 3x_2 \\ \max & f_3(\mathbf{x}) = & 2x_1 - x_2 \\ \text{s.t.} & & 3x_1 + x_2 \leq 30 \\ & & x_1 + x_2 \leq 20 \\ & & x_1 \leq 8 \\ & & x_1 \geq 0, x_2 \geq 0 \end{array}$$

The decomposition of the parametric (weight) diagram associated with the efficient vertices of the problem is displayed in the triangle.

What conclusions about the problem can be drawn from this decomposition?

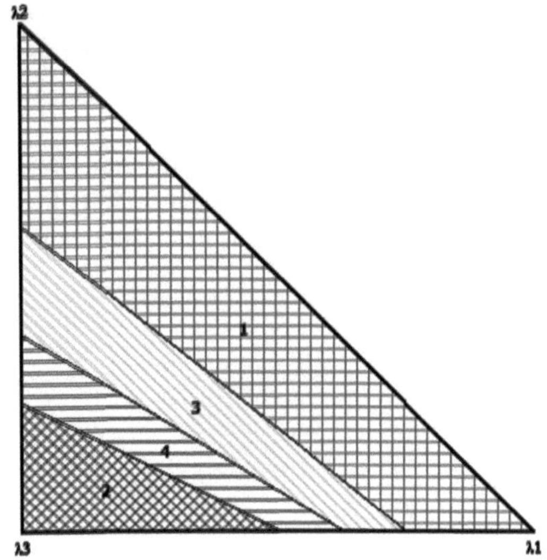


8. Consider the following MOLP problem:

$$\begin{aligned} \max f_1(\mathbf{x}) &= 3x_1 + x_2 \\ \max f_2(\mathbf{x}) &= x_1 + 2x_2 \\ \max f_3(\mathbf{x}) &= -x_1 + 2x_2 \\ \text{s.t.} \quad & -x_1 + x_2 \leq 2 \\ & x_1 + x_2 \leq 7 \\ & 0.5x_1 + x_2 \leq 5 \\ & x_1, x_2 \geq 0 \end{aligned}$$

(a) Represent graphically the set of efficient solutions.

(b) Suppose one wants to find the nondominated solution that minimizes the distance to the ideal solution by using a weighted Chebyshev metric. Formulate this problem, knowing that the ideal solution is $\mathbf{z}^* = (21, 10, 6)$ and considering the following (non-normalized) weights: $\lambda_1 = 1, \lambda_2 = 2, \lambda_3 = 1$.

(c) Consider the following reference point in the objective space, $\mathbf{q} = (14, 8, 0)$, belonging to the interior of the feasible region. Solution ($\mathbf{x}^a$, $\mathbf{z}^a$), with $\mathbf{x}^a = (4.2, 2.8)$ and $\mathbf{z}^a = (15.4, 9.8, 1.4)$, is optimal to the following problem:

$$\begin{aligned} & \min v \\ & \text{s. t.} \quad \lambda_1(14 - 3x_1 - x_2) \leq v \\ & \qquad \lambda_2(8 - x_1 - 2x_2) \leq v \\ & \qquad \lambda_3(0 + x_1 - 2x_2) \leq v \\ & \qquad \mathbf{x} \in X (X \text{ is the feasible region defined above}) \end{aligned}$$

with $\lambda = (1, 1, 1)$.

Identify how the variation trends of the objective function values evolve, regarding $\mathbf{z}^a$, when the previous problem is solved, but considering $\lambda = (2, 1, 1)$.

9. Ten vertex nondominated solutions to a three objective LP with $f_k(\mathbf{x})$, $k = 1,2,3$, were calculated using the weighted-sum scalarization, for which the corresponding indifference regions on the weight space are displayed.

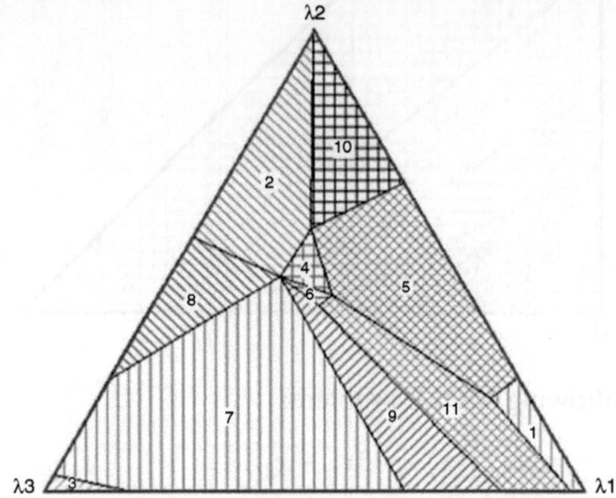


(a) Characterize all nondominated edges and faces using the vertices.
(b) Are there nondominated solutions that are alternative optima of any objective function?
(c) What are the nondominated vertices that can still be obtained with the weight constraints $\lambda_3 \geq \lambda_2 \geq \lambda_1$?
(d) What are the vertices of this three objective problem that are dominated in all 3 bi-objective problems that can be formed (i.e., $f_1(\mathbf{x})$ and $f_2(\mathbf{x})$, $f_1(\mathbf{x})$ and $f_3(\mathbf{x})$, $f_2(\mathbf{x})$ and $f_3(\mathbf{x})$)?
(e) Sketch the decomposition of the weight space for the problem with the objective functions $f_2(\mathbf{x})$ and $f_3(\mathbf{x})$. How do you classify solution 10 in this problem?

# Chapter 4
# Interactive Methods in Multiobjective Linear Programming

## 4.1 Introduction

In multiobjective programming problems, the methods dedicated to the generation of the whole set of nondominated solutions are in most cases inadequate from a practical point of view. In general, in real world models, the computational burden required for computing the entire set of nondominated solutions is too high. Moreover, proposing hundreds or thousands of solutions to a decision maker (DM) is not useful for the exploitation of results in practice, even limiting the computation to a subset of the nondominated solutions, for instance vertices of the feasible region in MOLP problems.

The use of utility function methods also does not seem the most adequate approach. In this case, although the model includes more than one objective function, their aggregation is made a priori, without further intervention of the DM after the definition of the utility function.

In our opinion, interactive methods enabling the progressive articulation of the DM's preferences are, in general, the most appropriate for decision support based on MOP models. Interactivity implies a succession of computation and dialogue phases. After each computation phase, one (or several) nondominated solution(s) is (are) proposed to the DM. He/she reacts providing the necessary information to start a new computation phase or deciding to stop the procedure, namely considering that sufficient information has been gathered to support the choice of a compromise solution.

The procedure for articulating the computation and dialogue phases allows classifying the interactive methods into two broad categories. The first encompasses interactive methods that constitute an evolution of the utility function methods, that is, methods where it is assumed that a DM's implicit utility function exists but he/she is not able to make it explicit. In this case, the role of the interactive protocol is essentially to discover the optimum (or an approximation



C. Henggeler Antunes et al., *Multiobjective Linear and Integer Programming*,
EURO Advanced Tutorials on Operational Research,
DOI 10.1007/978-3-319-28746-1_4

of it) of the implicit utility function, through a dialogue scheme. In opposition to this vision, free search methods do not have the purpose of converging to the optimum of any utility function, even an implicit one. Instead, the interactive protocol has an essential purpose, i.e. building progressively the DM's preferences in the nondominated solution set and not following a completely structured approach (Vincke 1992). Using the Feyerabend's (1975) terminology, a *guided exchange* underlying the first attitude is replaced by an *open exchange*, which is typified in the following way: "*An open exchange is guided by a pragmatic philosophy. The tradition adopted by the parties is unspecified in the beginning and develops as the exchange proceeds. The participants get immersed into each other's ways of thinking, feeling, perceiving to such an extent their ideas, perception, world views may be entirely changed*". These ideas are associated with group decision procedures; however, in our opinion, they are also well suited to interactive methods for multiobjective decision support based on a learning paradigm. In fact, in this type of approaches there is no place for mathematical convergence; the procedure stops when one or several satisfactory nondominated solution(s) is(are) obtained. In many cases, it is even excessive to call them methods, being essentially interactive computational environments, particularly adapted to the inclusion in decision support systems dedicated to a progressive and selective search for nondominated solutions in MOLP models.

In the illustrative study of the iMOLPe software package in Chap. 5, these issues are highlighted by offering several methods or procedures, since there is no method (or procedure) that performs best in all circumstances. Although iMOLPe can be used with different underlying strategies, it is our conviction that the search should be initiated by computing a set of, as much as possible, well distributed nondominated vertices. For this purpose, the optimization of weighted-sums of the objective functions seems adequate. In a second phase, the scope of the search can be delimited through the introduction of additional constraints on objective function values, according to the DM's indications in face of the information gathered. New solutions can then be computed in a reduced search region by combining, for instance, the optimization of weighted-sums of the objective functions with the minimization of a Chebyshev distance to a reference point.

A possible classification of interactive methods is based on the strategy for narrowing the scope of the search, the scalarizing problem used to (temporarily) aggregate the objective functions, and the flexibility offered to the DM's intervention to input his/her preference information (which in turn determines the reduction of the scope of the search). According to the progressive reduction of the search space, the following techniques may be distinguished: reduction of the feasible region, reduction of the weight space, reduction of the criterion cone (the cone spanned by the objective function gradients) and directional search. The three main scalarizing techniques used to compute nondominated solutions are the ones presented in Chap. 3: optimizing one of the objective functions and transforming the remaining into constraints, optimizing a weighted-sum of the objective functions, minimizing a distance/achievement function to a reference point. Regarding the flexibility offered to the DM to express his/her preferences, the interactive approaches may be classified according to their more or less rigid structure, i.e. the DM may intervene to require a certain operation at any point in the process or he/she needs to follow a predetermined sequence of steps.

In this chapter, five multiobjective interactive methods for MOLP problems, which are representative of different strategies of search, are described and illustrated using several examples. These methods are discussed having in mind the two broad categories previously mentioned, the procedures for the computation of nondominated solutions and strategies for reducing the scope of the search.

## 4.2 STEM Method

### 4.2.1 General Description

The *Step Method* (STEM) developed by Benayoun et al. (1971) is an interactive method that progressively reduces the feasible region.

In each computation phase a compromise solution is computed by minimizing a weighted Chebyshev distance to the ideal solution. If the objective function values are considered satisfactory, the procedure stops; otherwise, the DM should specify the amount that he/she is willing to sacrifice (relax) in the objective function whose value he/she considers satisfactory, in order to improve the objectives whose values are not yet acceptable. The feasible region is then progressively reduced, through limitations on the objective function values based on the relaxation amounts specified by the DM, and the procedure proceeds.

### 4.2.2 STEM Algorithm

**Step 1**
The objective functions are individually optimized to build the *pay-off* table.

Set $h = 1$ (iteration counter).

**Step 2**
Compute the weights $\beta_k$, using the values of the pay-off table. These weights are used in the computation phase. Their purpose is to take into account the orders of magnitude and the range of the objective function values in the computation of the nondominated solution that minimizes a weighted Chebyshev distance to the ideal solution.

$$\beta_k = \begin{cases} \underbrace{\dfrac{z_k^* - n_k}{z_k^*}}_{(A)} \underbrace{\left[\sum_{j=1}^{n} (c_{kj})^2\right]^{-\frac{1}{2}}}_{(B)} & \text{if} \quad z_k^* > 0 \\ \dfrac{n_k - z_k^*}{n_k} \left[\sum_{j=1}^{n} (c_{kj})^2\right]^{-\frac{1}{2}} & \text{if} \quad z_k^* \le 0 \end{cases} \quad k = 1, \ldots, p \tag{4.1}$$

where $n_k$ is the lowest value of the $k$ column of the pay-off table. Note that $n_k$ is an approximation to the minimum (worst value) of the objective function $f_k(\mathbf{x})$ in the feasible region. The term A in (4.1) privileges objective functions with higher relative variations in the nondominated region. The term B is a normalization factor concerning the objective function gradients, using the $L_2$ norm.

**Step 3**
The set $R$ includes the indices of the objective functions relaxed until the current iteration. In the first iteration $R = \varnothing$ and $X^{(1)} \equiv X$, where $X$ designates the feasible region.

The weights used in the weighted metric $L_\infty$, for the current iteration ($h$), are:

$$\alpha_k^{(h)} = \begin{cases} 0 & \text{if} \quad k \in R \\ \dfrac{\beta_k}{\sum_{i=1}^{p} \beta_i} & \text{if} \quad k \notin R \end{cases}$$

Note that the weights $\alpha_k^{(h)}$ corresponding to the objective functions relaxed until iteration $h$ are set to zero.

The weights $\alpha_k^{(h)}$ are normalized by making $\sum_{i=1}^{p} \alpha_i^{(h)} = 1$:

$$\alpha_k^{(h)} \leftarrow \frac{\alpha_k^{(h)}}{\sum_{i=1}^{p} \alpha_i^{(h)}} \qquad k = 1, \ldots, p$$

**Step 4**
In the computation phase, the linear problem that minimizes the weighted Chebyshev distance to the ideal solution is solved:

$$\begin{aligned} &\min v \\ &\text{s.t. } v \geq \alpha_k^{(h)} \left( z_k^* - \mathbf{c}_k \mathbf{x} \right), \quad 1 \leq k \leq p \\ &\qquad \mathbf{x} \in X^{(h)} \\ &\qquad v \geq 0 \end{aligned}$$

In the dialogue phase, the solution $\mathbf{z}^{(h)} = f\left(\mathbf{x}^{(h)}\right) = \mathbf{c}_k \mathbf{x}^{(h)}$, resulting from the resolution of the problem in iteration $h$, is presented to the DM. $\mathbf{x}^{(h)}$ is the point of the reduced feasible region $X^{(h)}$ corresponding to the point $\mathbf{z}^{(h)}$ closer to $\mathbf{z}^*$, according to the weighted Chebyshev metric.

**Step 5**
If the DM considers this solution satisfactory or $h = p$, the procedure stops with $(\mathbf{x}^{(h)}, \mathbf{z}^{(h)})$ as the final solution.

Otherwise, the DM is asked to indicate which objective function $f_i(\mathbf{x})$ ($R \leftarrow R \cup \{i\}$) he/she is willing to sacrifice, and the maximum amount $\Delta_i$ to be relaxed, in order to try to improve the functions whose values he/she did not yet consider satisfactory.

Set $h \leftarrow h + 1$.

**Step 6**
Based on the information gathered in the dialogue phase, the new computation phase is prepared by building the new reduced feasible region through the imposition of constraints on the objective function values. The reduced feasible region will then incorporate the constraints:

$f_i(\mathbf{x}) = \mathbf{c}_i \mathbf{x} \geq z_i^{(h)} - \Delta_i$ (corresponding to the objective function relaxed in iteration $h$)
$f_k(\mathbf{x}) = \mathbf{c}_k \mathbf{x} \geq z_k^{(h)}, \; k \neq i$

Return to step 3.

The working mechanism of the STEM method is displayed in the block diagram in Fig. 4.1.

### 4.2.3 *Final Comments*

Although in the version originally presented by the authors of the method each objective function can only be relaxed once and in a given iteration just one objective function can be relaxed, nothing prevents the elimination of these limitations to make the method more flexible. Nevertheless, changing the original version in that way the algorithm loses one of the essential characteristics claimed by the authors, which is converging to a final solution in a maximum number of $p$ iterations. This characteristic was relevant by the time this method was developed, not only due to the computational limitations but also because the idea that

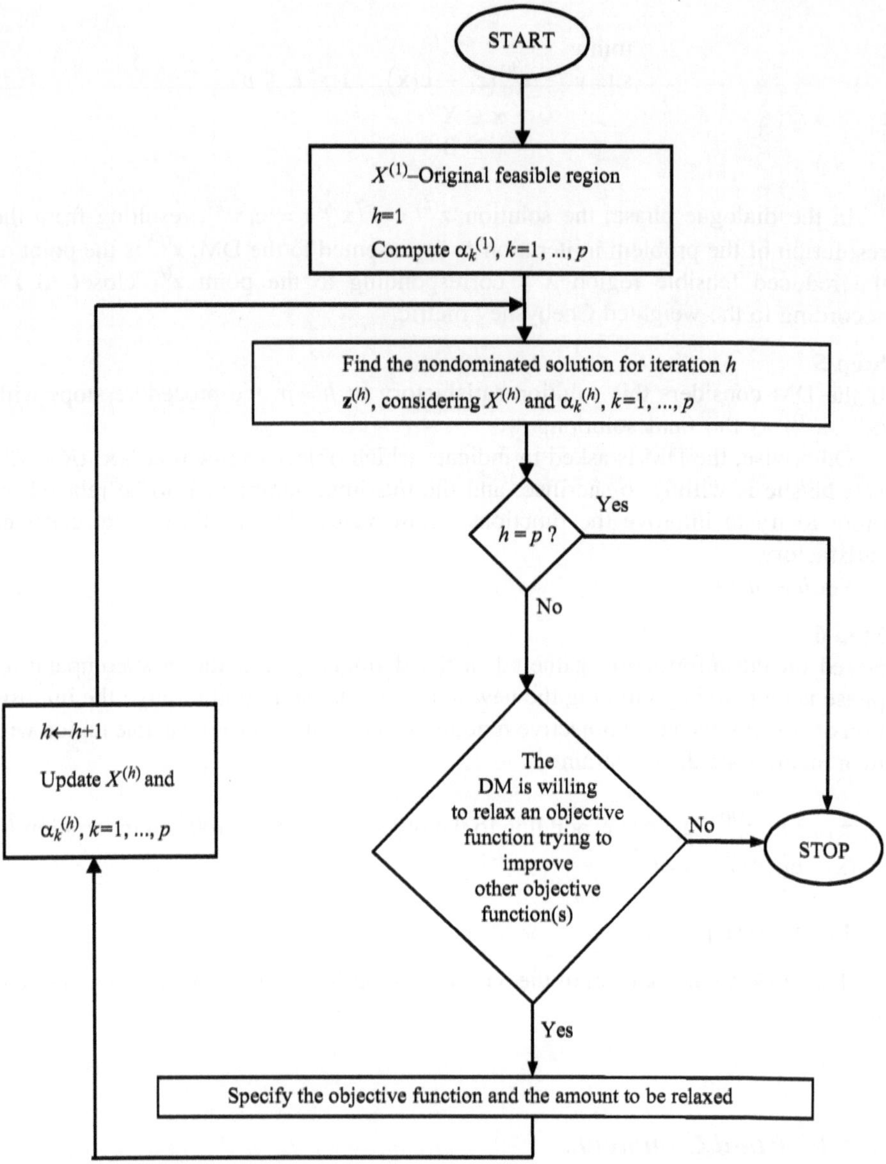


**Fig. 4.1** STEM block diagram

interactive methods should converge to a certain optimal solution was prevailing. As it was seen in the introduction of this chapter, methods with these characteristics can be integrated in a flexible and *learning oriented* manner during the decision support process. By allowing more than one objective function to be relaxed in each iteration and to review the relaxed amount of the objective functions in subsequent iterations, a learning oriented procedure is built, which stops when the DM considers having obtained a satisfactory solution. In this way the imposition of irrevocable decisions is avoided and, in particular, the obligation of establishing in a rigid manner (without the possibility of re-evaluation) the amounts being relaxed can be made smoother.

This method is very easy to implement with a standard linear programming code, and the search is not limited to the nondominated vertices of the feasible region (due to the procedure used for computing nondominated solutions). Note that, in the version herein presented, it is possible to obtain weakly nondominated solutions because the non-augmented weighted Chebyshev metric is used in Step 4. This issue, which is illustrated in Fig. 4.7, can be overcome using the augmented weighted Chebyshev metric (cf. Chap. 3).

### 4.2.4 *Illustrative Example of the STEM Method*

Consider the following problem, with two objective functions being maximized:

$$\begin{array}{ll} \max & z_1 = f_1(\mathbf{x}) = 3x_1 + x_2 \\ \max & z_2 = f_2(\mathbf{x}) = x_1 + 4x_2 \\ s.t. & -x_1 + x_2 \leq 2 \\ & \quad x_1 + x_2 \leq 7 \\ & \quad x_1 + 2x_2 \leq 10 \\ & \quad x_1, x_2 \geq 0 \end{array}$$

Since this problem has two decision variables and two objective functions, the decision variable space (Fig. 4.2) and the objective function space (Fig. 4.3) can be easily visualized.

The STEM method starts by individually optimizing each objective function in the original feasible region of the multiobjective problem. With this information the *pay-off* table is built. In this problem it is possible to see graphically that A (A′) is the solution that optimizes $f_1(\mathbf{x})$ and C (C′) optimizes $f_2(\mathbf{x})$. Thus, the *pay-off* table is:

| | $z_1$ | $z_2$ |
|---|---|---|
| A′ | **21** | 7 |
| C′ | 10 | **18** |

The ideal solution is $\mathbf{z}^* = (z_1^*, z_2^*) = (21, 18)$.

The weights $\beta_k$ are:

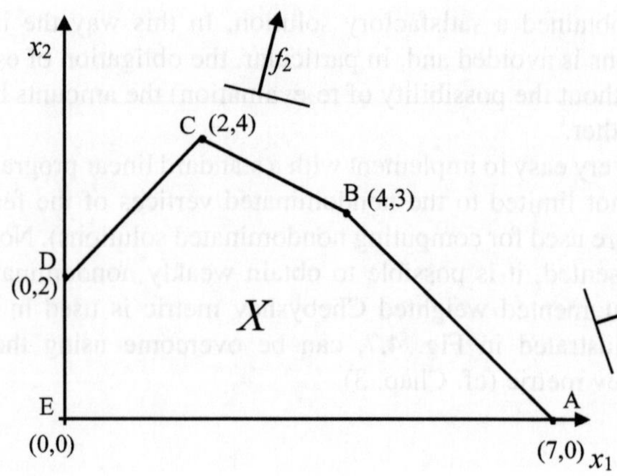


**Fig. 4.2** Decision space

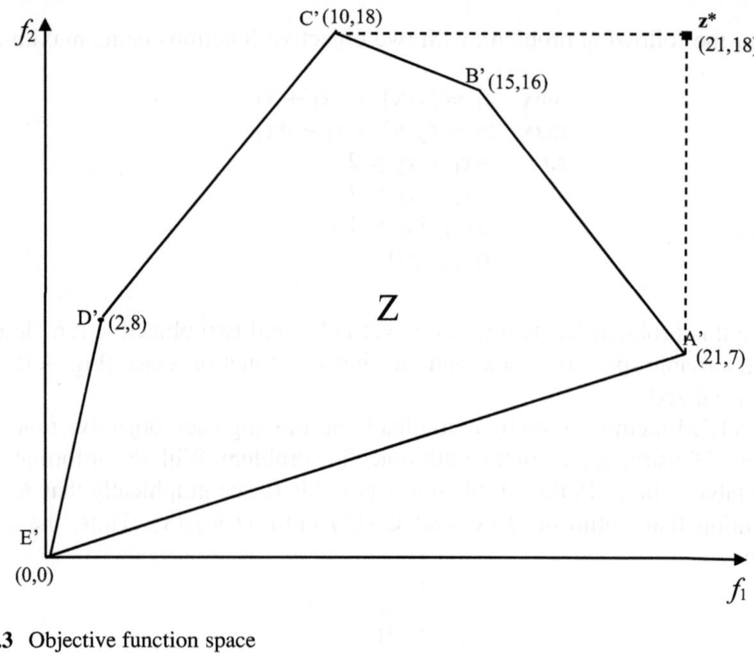


**Fig. 4.3** Objective function space

$$\beta_1 = \frac{21-10}{21}\left(\frac{1}{\sqrt{3^2+1^2}}\right) = 0.1656$$

$$\beta_2 = \frac{18-7}{18}\left(\frac{1}{\sqrt{1^2+4^2}}\right) = 0.1482$$

**First Iteration**

Let $h$ be the iteration counter. Set $h = 1$.

The feasible region of the first iteration $X^{(1)}$ is the initial feasible region $X$.

$R = \emptyset$ (set of the indices of the functions relaxed so far).
The weights $\alpha_k$ are:

$$\alpha_1^{(1)} = \frac{\beta_1}{\beta_1 + \beta_2} = 0.528$$

$$\alpha_2^{(1)} = \frac{\beta_2}{\beta_1 + \beta_2} = 0.472$$

In order to find the solution that minimizes the weighted Chebyshev distance to the ideal solution, the following scalarizing problem is solved:

$$\begin{aligned} &\min v \\ &\text{s.t.} \quad 0.528(21 - (3x_1 + x_2)) \leq v \\ &\qquad 0.472(18 - (x_1 + 4x_2)) \leq v \\ &\qquad \left.\begin{aligned} -x_1 + x_2 &\leq 2 \\ x_1 + x_2 &\leq 7 \\ x_1 + 2x_2 &\leq 10 \\ x_1 &\geq 0 \\ x_2 &\geq 0 \end{aligned}\right\} \text{original feasible region } X \\ &\qquad v \geq 0 \end{aligned}$$

This problem is equivalent to:

$$\begin{aligned} &\min v \\ &\text{s.t.} \quad 1.584x_1 + 0.528x_2 + v \geq 11.088 \\ &\qquad 0.472x_1 + 1.888x_2 + v \geq 8.5 \\ &\qquad \mathbf{x} \in X \\ &\qquad v \geq 0 \end{aligned}$$

The solution to this problem is $\mathbf{x}^{(1)} = (4.9,\ 2.1)$, $\mathbf{z}^{(1)} = (16.8, 13.3)$— see Fig. 4.4.

Suppose that the DM considers the value of $f_1(\mathbf{x})$ satisfactory in this solution and that he/she admits to worsen it by an amount not higher than 2.8, in order to try to improve the value of $f_2(\mathbf{x})$.

Then, $R = \{1\}$ and the amount being relaxed is $\Delta_1 = 2.8$.

The feasible region of the next iteration, $X^{(2)}$, is defined adding to the original constraints the following constraints: $f_1(\mathbf{x}) \geq z_1^{(1)} - \Delta_1$ and $f_2(\mathbf{x}) \geq z_2^{(1)}$, corresponding to $f_1(\mathbf{x}) \geq 16.8 - 2.8$ and $f_2(\mathbf{x}) \geq 13.3$, respectively.

Then, $X^{(2)}$ is defined by:

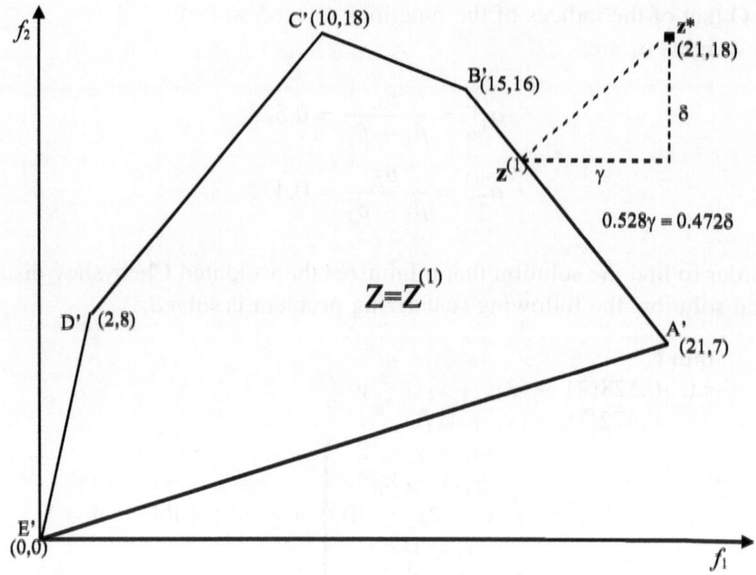


**Fig. 4.4** $z^{(1)}$ is the nondominated solution obtained in the initial iteration

$$
\begin{aligned}
-x_1 + x_2 &\leq 2 \\
x_1 + x_2 &\leq 7 \\
x_1 + 2x_2 &\leq 10 \\
3x_1 + x_2 &\geq 14 \\
x_1 + 4x_2 &\geq 13.3 \\
x_1, x_2 &\geq 0
\end{aligned}
$$

Figure 4.5 shows the reduced feasible region in the decision space.

**Second Iteration**

$h = 2;$
$\alpha_1^{(2)} = 0$ (because $1 \in R$, that is, $f_1(\mathbf{x})$ has already been relaxed)
$\alpha_2^{(2)} = 1$

Solve the following problem:

$$
\begin{aligned}
&\min v \\
&\text{s.t. } 18 - (x_1 + 4x_2) \leq v \\
&\qquad \mathbf{x} \in X^{(2)} \\
&\qquad v \geq 0
\end{aligned}
$$

The solution to this problem is $\mathbf{x}^{(2)} = (3.6, 3.2)$ (see Fig. 4.5), with $\mathbf{z}^{(2)} = (14, 16.4)$ (see Fig. 4.6).

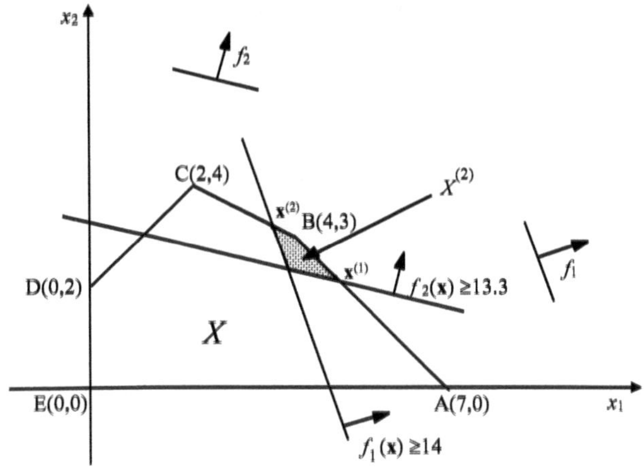


**Fig. 4.5** Reduced feasible region in the decision space after the first iteration

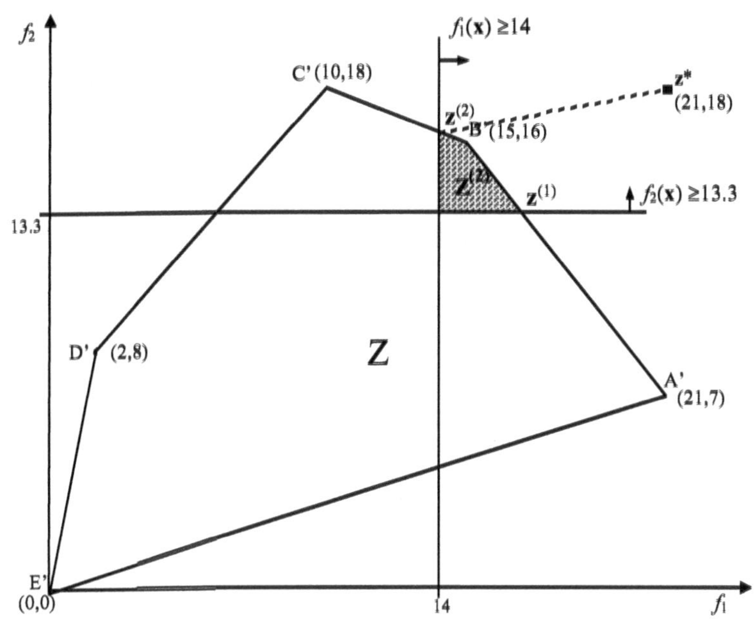


**Fig. 4.6** Reduced feasible region in the objective function space

The procedure ends with ($\mathbf{x}^{(2)}$, $\mathbf{z}^{(2)}$) as the final solution, since this problem has two objective functions only and $h = 2$.

Suppose that the problem was slightly changed, with the second objective function being $z_2 \equiv f_2(\mathbf{x}) = x_1 + 2x_2$. Note that, in this case, alternative optimal solutions to $f_2(\mathbf{x})$ are located on the edge CB (Fig. 4.7) and all solutions on this edge are just weakly efficient, except B that is strictly efficient.

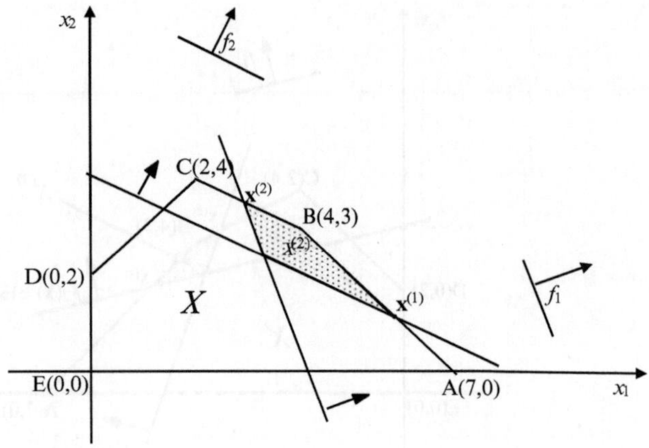


**Fig. 4.7** Weakly efficient solutions

In the STEM method, this change would lead, after the first iteration, to the situation presented in Fig. 4.7. As there are alternative optima to $f_2(\mathbf{x})$ in $X$, in the second iteration it is possible to obtain $\mathbf{x}^{(2)}$ that is just a weakly efficient solution. This can be avoided, i.e. a strictly efficient solution can be enforced by "perturbing" the Chebyshev scalarizing function as described in Chap. 3, so that the efficient solution B is obtained.

## 4.3 Zionts and Wallenius Method

### 4.3.1 Introduction

The Zionts and Wallenius (1976, 1983) method progressively reduces the weight space, according to the DM's preferences in each interaction. These preferences are expressed by answers regarding pairwise comparisons between solutions and judgments about the marginal rates of variation of the objective functions associated with the edges emanating from the current (basic) solution and leading to adjacent nondominated (basic) solutions. In each computation phase a weighted-sum of the objective functions is optimized.

The method introduces constraints on the weight space derived from the answers given by the DM, thus progressively reducing the feasible domain for selecting a new weight vector. The procedure stops when the weight space is reduced to a sufficiently small region, so that it is possible to identify a final solution, or when the information of preferences expressed by the DM indicates that the current solution is the most interesting one. Then, it is assumed that the process *converges* to the optimum of the DM's implicit utility function or, more precisely, to the nondominated vertex that leads to the highest value of that function. It is assumed that the DM's answers, in the dialogue phases, are coherent with that implicit utility function, although in the case of inconsistency being detected (for instance, revealed by the reduced weight space becoming empty) there is the possibility of eliminating the oldest constraints until the weight space is non-empty again, thus allowing to proceed the search.

### 4.3.2 *Zionts and Wallenius Algorithm*

The algorithm proposed in Zionts and Wallenius (1976,1983) is presented below. A detailed study of this algorithm can be found in Steuer (1986).

**Step 1**

A weight vector $\boldsymbol{\lambda}^{(1)} \in \Lambda = \left\{ \boldsymbol{\lambda} \in \mathbb{R}^p : \lambda_k > 0, \sum_{k=1}^{p} \lambda_k = 1 \right\}$ (original weight space) is chosen, and the following linear programming problem is solved:

$$\begin{array}{ll} \max & \lambda^{(1)}\mathbf{C}\mathbf{x} \\ \text{s.t.} & \mathbf{x} \in X \end{array}$$

An efficient basic solution $\mathbf{x}^{(1)}$ is obtained and the corresponding image in the objective function space is the nondominated vector $\mathbf{z}^{(1)} = \mathbf{C}\ \mathbf{x}^{(1)}$.

Although any initial weight vector, $\boldsymbol{\lambda}^{(1)} \in \Lambda$ can be chosen, in general, the central point of $\Lambda : \lambda^{(1)} = \left(\frac{1}{p}, \frac{1}{p}, \ldots, \frac{1}{p}\right)$ is adopted.

Set $h = 1$ (iteration counter).

**Step 2**

The set of efficient nonbasic variables is split into two sub-sets, $A$ and $B$:

- $A$ designates the set of efficient nonbasic variables, which becoming basic lead to efficient vertices attainable by optimizing weighted-sum functions with $\lambda \in \Lambda^{(h)}$. In the first iteration $\Lambda^{(1)} = \Lambda$.
- $B$ designates the set of efficient nonbasic variables that are not in $A$.

In order to find vertices adjacent to the current solution $\mathbf{x}^{(h)}$, it is necessary to know the nonbasic variables that are efficient (making basic an efficient nonbasic variable leads to an adjacent efficient vertex and the edge that connects the two vertices also consists of efficient solutions). Zionts and Wallenius (1980) proposed a routine for classifying nonbasic variables as to whether they are efficient or not. This routine allows to determine the efficient nonbasic variables with respect to the basis associated with the current solution $\mathbf{x}^{(h)}$, as well as to build the sets $A$ and $B$.

Let the set $I = A$.

**Step 3**

The efficient basic solutions $\mathbf{x}^{adj}$ adjacent to $\mathbf{x}^{(h)}$ are computed by making basic each efficient nonbasic variable belonging to $I$. This process is implemented through a pivot operation in the multiobjective simplex tableau.

The vectors of the objective function values $\mathbf{z}^{adj}$ that are not sufficiently distinct from $\mathbf{z}^{(h)}$ are temporarily not considered. Two points in the objective function space, $\mathbf{z}^{adj}$ and $\mathbf{z}^{(h)}$, are considered sufficiently distinct if there is a minimal difference of 10 % between them in at least one of the objective function values (this procedure avoids to compare very similar solutions concerning the objective function values). This threshold can change according to the context of the problem. Here, we have adopted the percentage of 10 % proposed by Steuer (1986).

For each point $\mathbf{z}^{adj}$, sufficiently distinct from $\mathbf{z}^{(h)}$, the DM is asked to make pairwise comparisons and express his/her preferences according to:

1. Yes — if the DM prefers the solution $\mathbf{z}^{adj}$ adjacent to $\mathbf{z}^{(h)}$.
2. No — if the DM prefers $\mathbf{z}^{(h)}$ to $\mathbf{z}^{adj}$.
3. Does not know — if the DM is not able to express a preference.

The pairwise comparisons can be made for all the $\mathbf{z}^{adj}$ sufficiently distinct from $\mathbf{z}^{(h)}$, or just until obtaining an affirmative answer (that is, until a $\mathbf{z}^{adj}$ that the DM prefers to $\mathbf{z}^{(h)}$ is identified).

If the DM prefers a particular adjacent solution $\mathbf{z}^{a}$ instead of the current solution $\mathbf{z}^{(h)}$, then $\mathbf{z}^{a}$ is selected and the algorithm goes to step 7. Note that more than one solution $\mathbf{z}^{a}$ may exist.

If there is no solution $\mathbf{z}^{a}$, i.e. the DM prefers $\mathbf{z}^{(h)}$, then the algorithm goes to step 4.

**Step 4**

The algorithm reaches this step because there is no adjacent solution (sufficiently distinct from the current one) that is preferred by the DM. At this stage, the DM has the possibility of evaluating the interest of the marginal rates of variation of the objective functions (*trade-offs*) along the nondominated edges leading to the adjacent vertices not sufficiently distinct from the current one.

Regarding the set $I$, the vectors representing the marginal rates of variation of the objective functions along the efficient edges are generated for all efficient nonbasic variables the DM was not asked about in step 3 (i.e., the ones leading to adjacent efficient solutions not sufficiently distinct as defined above). Note that the marginal rates of variation through non-efficient edges are not investigated, although they may lead to efficient vertices (see de Samblanckx et al. 1982).

Each vector of marginal rates of variation of the objective functions ($\mathbf{w} = \mathbf{W}_{.j}$, where $x_j$ is an efficient nonbasic variable[1]) is evaluated by the DM expressing his/her preferences as follows:

1. Yes — if the DM accepts the marginal rate of variation.
2. No — if the DM does not accept the marginal rate of variation.
3. Does not know — if the DM is not able to express a preference.

[1] In the multiobjective simplex tableau, the column of **W** corresponding to a nonbasic variable indicates the objective function variations per unit of that nonbasic variable when it becomes basic. It should be noted that, as introduced in Chap. 3, the positive values of the reduced cost matrix **W** correspond to worsening the corresponding objective function values.

If the DM accepts the marginal rate of variation associated with an unbounded edge, the method stops with an unbounded solution.

If the DM accepts at least one of the marginal rates of variation corresponding to a bounded edge, then the algorithm goes to step 7; otherwise, it proceeds to step 5.

**Step 5**

The algorithm reaches this step because there is no adjacent solution (sufficiently distinct from the current one) preferred by the DM in comparison with the current one, and because the marginal rates of variation along the efficient edges leading to adjacent efficient vertices (not sufficiently distinct from the current one) are not considered interesting.

Then, the DM has the opportunity of identifying the edges which he/she considers to be interesting to go through, although the extreme points (vertices) of these edges have not been preferred to the current solution according to the answers given in the pairwise comparisons in step 3. That is, regarding the set $I$, the DM is asked to specify if he/she accepts any marginal rate of variation associated with the efficient edges leading to efficient vertices (which were not preferred in step 3) adjacent to the current solution.

Each vector of marginal rates of variation in these conditions is evaluated by the DM, expressing his/her preferences as follows:

1. Yes — if the DM accepts the marginal rate of variation.
2. No — if the DM does not accept the marginal rate of variation.
3. Does not know — if the DM is not able to express a preference.

If the DM accepts at least one of these marginal rates of variation, the algorithm proceeds to step 7; otherwise, it goes to step 6.

**Step 6**

The algorithm reaches this step because there exists neither an adjacent solution preferred to the current one, nor interesting marginal rates of variation of the objective functions along the edges emanating from the current solution.

If $I = A$, this means that all nonbasic variables have been examined, leading to all the efficient solutions achievable in the current weight space, $\Lambda^{(h)}$.

Then, the algorithm makes $I = B$, in order to examine the other sub-set of efficient nonbasic variables, and returns to step 3. In this way, all efficient nonbasic variables, corresponding to all efficient edges emanating from $\mathbf{x}^{(h)}$, are examined even if they lead to efficient vertices impossible to reach in $\Lambda^{(h)}$ (that is, not coherent with the DM's previous answers).

If $I = B$, it means that all the adjacent vertices generated from the current solution (by making basic an efficient nonbasic variable) and all the marginal rates of variation along the efficient edges having origin in the current solution have already been examined. If the answers regarding these questions are "no" or "does not know", the final solution is $\mathbf{x}^{(h)}$ and its image in the objective function space is $\mathbf{z}^{(h)}$. In this case the algorithm stops.

**Step 7**

The algorithm reaches this step because there is at least a positive answer in the pairwise comparisons (in step 3) or in the evaluation of the marginal rates of variation (in steps 4 or 5).

Constraints are introduced in the weight space, based on the DM's answers regarding the pairwise comparisons (in step 3) and the evaluation of marginal rates of variation (in steps 4 or 5).

Each pairwise comparison, between $\mathbf{z}^{adj}$ and $\mathbf{z}^{(h)}$ made by the DM in step 3, generates a constraint on the weight space (except for the answers "does not know", which do not generate constraints), as follows:

$$\boldsymbol{\lambda}\left(\mathbf{z}^{adj} - \mathbf{z}^{(h)}\right) \geq \varepsilon \quad \text{for each affirmative answer (that is, the DM prefers } \mathbf{z}^{adj} \text{ to } \mathbf{z}^{(h)}\text{),}$$

$$\boldsymbol{\lambda}\left(\mathbf{z}^{adj} - \mathbf{z}^{(h)}\right) \leq -\varepsilon \quad \text{for each negative answer (that is, the DM prefers } \mathbf{z}^{(h)} \text{ to } \mathbf{z}^{adj}\text{),}$$

where $\varepsilon$ is a very small positive value. These inequalities are used instead of strict inequalities with a zero right hand side due to numerical reasons.

Each evaluation of the marginal rates of variation (steps 4 and 5) leads to a constraint on the weight space, as follows:

$$\boldsymbol{\lambda}\mathbf{w} \geq \varepsilon \quad \text{for each negative answer (that is, the DM does not accept the marginal rate of variation),}$$

$$\boldsymbol{\lambda}\mathbf{w} \leq -\varepsilon \quad \text{for each affirmative answer (that is, the DM accepts the marginal rate of variation).}$$

The reduced weight space $\Lambda^{(h+1)}$ is built from $\Lambda^{(h)}$, by adding the constraints herein generated.

**Step 8**

A point $\boldsymbol{\lambda}^{(h+1)} \in \Lambda^{(h+1)}$ is determined. When $\Lambda^{(h+1)} = \varnothing$ it is not possible to determine $\lambda^{(h+1)}$ (it may happen that due to the changes in the DM's preferences inconsistent answers lead to $\Lambda^{(h+1)} = \varnothing$). In this case we should start by eliminating, from the set of binding constraints delimiting the reduced weight space, the oldest ones until $\Lambda^{(h+1)}$ is non-empty.

In order to determine a "central" point $\boldsymbol{\lambda}^{(h+1)} \in \Lambda^{(h+1)}$ a linear programming problem is solved, such that $\boldsymbol{\lambda}^{(h+1)}$ is the point that maximizes the smallest deviation (*slack*) to the frontier defined by the constraints delimiting the reduced weight space $\Lambda^{(h+1)}$ (see example below).

**Step 9**
A weighted-sum linear programming problem is solved using $\boldsymbol{\lambda}^{(h+1)}$:

$$\begin{array}{ll} \max & \boldsymbol{\lambda}^{(h+1)}\mathbf{C}\mathbf{x} \\ \text{s.t.} & \mathbf{x} \in X \end{array}$$

A new efficient solution $\mathbf{x}^{(b)}$ is obtained, with image $\mathbf{z}^{(b)}$ in the objective function space.

**Step 10**
In the present situation, the DM is asked to express his/her preferences between the new nondominated solution $\mathbf{z}^{(b)}$ (obtained in step 9) and $\mathbf{z}^{(a)}$ (preferred to $\mathbf{z}^{(h)}$, in the pairwise comparisons performed in step 3) if this exists. Note that, if there is more than one $\mathbf{z}^{(a)}$ selected in step 3, it is then necessary to choose the preferred $\mathbf{z}^{(a)}$ among them.

If $\mathbf{z}^{(a)}$ exists and the DM is able to choose between $\mathbf{z}^{(a)}$ and $\mathbf{z}^{(b)}$, the constraint corresponding to this preference is added to the weight space $\left(\boldsymbol{\lambda}\left(\mathbf{z}^{(b)} - \mathbf{z}^{(a)}\right) \geq \varepsilon \text{ or } \boldsymbol{\lambda}\left(\mathbf{z}^{(b)} - \mathbf{z}^{(a)}\right) \leq -\varepsilon\right)$, and the preferred solution is designated by $\mathbf{z}^{(h+1)}$. Go to step 11.

Otherwise, that is, if the DM is not able to express a preference between $\mathbf{z}^{(a)}$ and $\mathbf{z}^{(b)}$, then make $\mathbf{z}^{(h+1)} = \mathbf{z}^{(b)}$. Go to step 11.

If $\mathbf{z}^{(a)}$ does not exist, i.e. no solution was preferred to $\mathbf{z}^{(h)}$, in the pairwise comparisons in step 3, two situations can occur:

(i) $\mathbf{z}^{(b)}$ is preferred to $\mathbf{z}^{(h)}$. The constraint corresponding to this preference $\boldsymbol{\lambda}\left(\mathbf{z}^{(b)} - \mathbf{z}^{(h)}\right) \geq \varepsilon$ is added to the weight space and $\mathbf{z}^{(b)}$ is the new current solution, that is: $\mathbf{z}^{(h+1)} = \mathbf{z}^{(b)}$. Go to step 11.

(ii) $\mathbf{z}^{(h)}$ is the preferred solution. In this case, the method ends with the current solution $\mathbf{x}^{(h)}$ (whose image in the objective function space is $\mathbf{z}^{(h)}$) as the final solution. Note that better solutions (according with the DM's implicit utility function) might exist on an efficient facet, since the method only computes the feasible region vertex solutions due to the type of scalarizing function used.

**Step 11**
In the case $\mathbf{x}^{(a)}$ and $\mathbf{z}^{(a)}$ exist, these are eliminated and the reduced weight space $\Lambda^{(h+1)}$ includes the new constraints that have been introduced in step 10.

Set $h \leftarrow h+1$, returning to step 2: a new iteration is made, restarting from the new current solution considering the reduced weight space.

The possibility of incoherence among the constraints in the weight space corresponding to the answers of the DM and the fact that the method, in these circumstances, eliminates the older constraints (avoiding that the reduced weight space becomes empty), led Ramesh et al. (1989) to propose a modification to the Zionts and Wallenius method. This modification aims to avoid loosing information

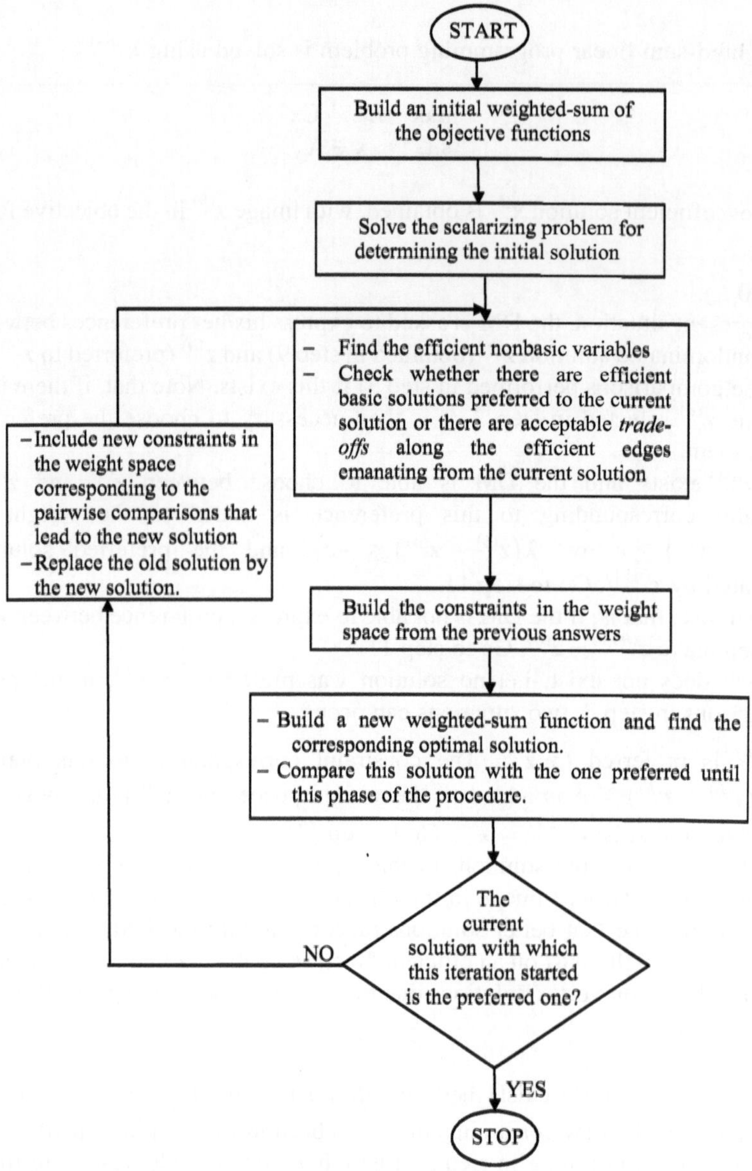


**Fig. 4.8** Block diagram of the Zionts and Wallenius method

about the DM's preferences expressed in the answers throughout the interactive process.

Figure 4.8 shows a simplified block diagram of the Zionts and Wallenius method.

### 4.3.3 Final Comments

This method is based on the existence of a DM's implicit utility function. The interactive procedure consists in the indirect search of its optimum (or an approximation of it) since the utility function is not explicitly known, through a dialogue protocol with the DM. Since the computation procedure only allows obtaining efficient basic solutions, the approximation found is restrained to this type of solutions. From the cognitive point of view, the questions based on pairwise comparison of solutions and *trade-offs* (marginal rates of variation of the objective functions) evaluation are not easy for the DM, even admitting that the procedure is facilitated by an analyst dealing with technical issues. In particular, concerning the *trade-offs* due to the unit activation of an efficient nonbasic variable, the DM is asked to accept or reject a small displacement along an edge without knowing its length, and hence the efficient vertex in its extremity.

The number of linear programs being solved is, in general, very high, making the method quite heavy from the computational point of view.

Finally, it is highlighted the fact that each answer of the DM (except when he/she avoids intervention by saying that he/she "does not know") corresponds to introduce a dichotomic constraint in the weight space. A wrong answer leads to excluding the entire region of the weight space that should be explored in subsequent phases of the method. This limitation is of concern, since the number of questions of this type is very high. In these circumstances, although it is possible to correct certain type of mistakes, the method requires the coherence of the DM's answers with his/her implicit utility function.

### 4.3.4 Illustrative Example of the Zionts and Wallenius Method

Consider the following MOLP problem with three objective functions to be maximized:

$$\begin{aligned}
\max \quad & z_1 = f_1(\mathbf{x}) = 3x_1 + x_2 + 2x_3 + x_4 \\
\max \quad & z_2 = f_2(\mathbf{x}) = x_1 - x_2 + 2x_3 + 4x_4 \\
\max \quad & z_3 = f_3(\mathbf{x}) = -x_1 + 5x_2 + x_3 + 2x_4 \\
\text{s.t.} \quad & \\
& 2x_1 + x_2 + 4x_3 + 3x_4 \leq 60 \\
& 3x_1 + 4x_2 + x_3 + 2x_4 \leq 60 \\
& x_1 + 2x_2 + 3x_3 + 4x_4 \leq 50 \\
& x_1, x_2, x_3, x_4 \geq 0
\end{aligned}$$

**First iteration**

Step 1. Let $h$ be the iteration counter: $h = 1$.

The first efficient basic solution is computed through the optimization of a weighted-sum of the objective functions. By default, the central weight vector of the weight space is considered: $\boldsymbol{\lambda}^{(1)} = (0.333, 0.333, 0.333)$. The problem to be solved is:

$$\begin{array}{ll} \max & 0.333(3x_1 + x_2 + 2x_3 + x_4) \\ & +0.333(x_1 - x_2 + 2x_3 + 4x_4) \\ & +0.333(-x_1 + 5x_2 + x_3 + 2x_4) \\ \text{s.t.} & 2x_1 + x_2 + 4x_3 + 3x_4 \leq 60 \\ & 3x_1 + 4x_2 + x_3 + 2x_4 \leq 60 \\ & x_1 + 2x_2 + 3x_3 + 4x_4 \leq 50 \\ & x_1, x_2, x_3, x_4 \geq 0 \end{array} \qquad (P_{\lambda^1})$$

The solution $\mathbf{x}^{(1)} = (0, 11.67, 0, 6.67)$ is optimal to $(P_{\lambda^1})$ and it is efficient to the multiobjective problem, having as image $\mathbf{z}^{(1)} = (18.33, 15.00, 71.67)$ in the objective function space. Note that $\mathbf{x}^{(1)}$ is not the only solution which optimizes $(P_{\lambda^1})$, as it can be seen in the corresponding simplex tableau.

The optimal simplex tableau of $(P_{\lambda^1})$ regarding $\mathbf{x}^{(1)}$ (where $x_5$, $x_6$ and $x_7$ are the slack variables associated with the constraints) is:

| | c | 1 | 1.667 | 1.667 | 2.333 | 0 | 0 | 0 | |
|---|---|---|---|---|---|---|---|---|---|
| $(\mathbf{c}_B)^T$ | $\mathbf{x}_B$ | $x_1$ | $x_2$ | $x_3$ | $x_4$ | $x_5$ | $x_6$ | $x_7$ | |
| 0 | $x_5$ | 1.667 | 0 | 1.667 | 0 | 1 | 0.167 | −0.833 | 28.333 |
| 1.667 | $x_2$ | 0.833 | 1 | −0.167 | 0 | 0 | 0.333 | −0.167 | 11.667 |
| 2.333 | $x_4$ | −0.167 | 0 | 0.833 | 1 | 0 | −0.167 | 0.333 | 6.667 |
| $z_j^1 - c_j^1$ | | −2.333 | 0 | −1.333 | 0 | 0 | 0.167 | 0.167 | 18.333 |
| $z_j^2 - c_j^2$ | | −2.5 | 0 | 1.5 | 0 | 0 | −1 | 1.5 | 15 |
| $z_j^3 - c_j^3$ | | 4.833 | 0 | −0.167 | 0 | 0 | 1.333 | −0.167 | 71.667 |
| $z_j^\lambda - c_j^\lambda$ | | 0 | 0 | 0 | 0 | 0 | 0.167 | 0.5 | |

The indifference region of solution 1 (Fig. 4.9) in the weight space is defined by:

$$\lambda \in \Lambda: \quad \begin{array}{ll} -2.333\lambda_1 - 2.5\lambda_2 + 4.833\lambda_3 \geq 0 \\ -1.333\lambda_1 + 1.5\lambda_2 - 0.167\lambda_3 \geq 0 \\ 0.167\lambda_1 - \lambda_2 + 1.333\lambda_3 \geq 0 \\ 0.167\lambda_1 + 1.5\lambda_2 - 0.167\lambda_3 \geq 0 \end{array}$$

with $\Lambda = \{\boldsymbol{\lambda} \in \mathbb{R}^3: \lambda_k > 0, k = 1,2,3, \sum_{k=1}^{3} \lambda_k = 1\}$.

The solution $(\mathbf{x}^{(1)}, \mathbf{z}^{(1)})$ is presented to the DM.

Step 2. The reduced cost matrix regarding the nonbasic variables $x_1, x_3, x_6$ and $x_7$ associated with solution $(\mathbf{x}^{(1)}, \mathbf{z}^{(1)})$ is:

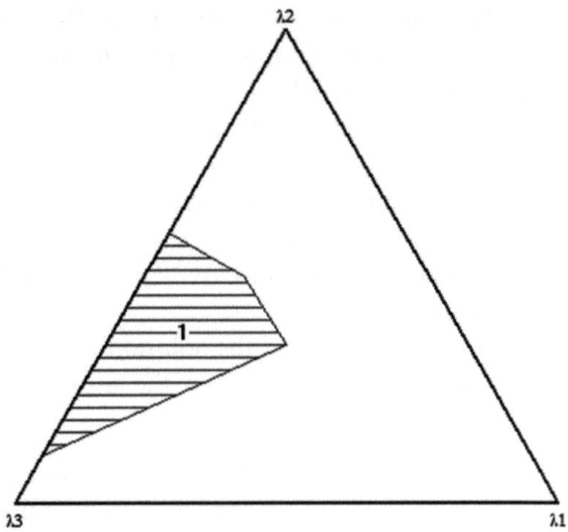


**Fig. 4.9** Indifference region of solution 1

$$\mathbf{W} = \begin{bmatrix} -2.333 & -1.333 & 0.167 & 0.167 \\ -2.5 & 1.5 & -1 & 1.5 \\ 4.833 & -0.167 & 1.333 & -0.167 \end{bmatrix}$$

The set of efficient nonbasic variables is divided into two sub-sets:

$A$ — efficient nonbasic variables that generate nondominated solutions reachable using weight vectors belonging to $\Lambda^{(1)}$;
$B$ — efficient nonbasic variables not belonging to $A$.

Since $\Lambda^{(1)} = \Lambda$ (because no weight constraint has been introduced yet), all the efficient nonbasic variables belong to $A$ and $B = \emptyset$.

In order to determine the efficient nonbasic variables, the Zionts-Wallenius routine (Steuer 1986, chap. 9) is applied to matrix $\mathbf{W}$.

After the application of this routine, it is concluded that, among $x_1, x_3, x_6$ and $x_7$, only $x_1, x_3, x_6$ are efficient (corresponding to the first, second and third columns of $\mathbf{W}$). In this case, the fact that $x_7$ is not efficient can be directly observed in matrix $\mathbf{W}$. Comparing the column of $x_3$ with the column of $x_7$, it is verified that only the first component is different (–1.333 for $x_3$ and 0.167 for $x_7$), meaning that making $x_3$ and $x_7$ basic variables has the same impact on $f_2(\mathbf{x})$ and $f_3(\mathbf{x})$. However, there is an increase of 1.333 in $f_1(\mathbf{x})$ for each unit of increase in $x_3$, while if $x_7$ becomes a basic variable there is a decrease of 0.167 for each unit it increases.

$$A = \{x_1, x_3, x_6\}$$

Set $I = A$.

Step 3. All the efficient solutions adjacent to $(\mathbf{x}^{(1)}, \mathbf{z}^{(1)})$ corresponding to the set $I$ are generated.

- $x_1$ becomes a basic variable and $\mathbf{z}^{adj1} = (51, 50, 4)$ is obtained. This solution is presented to the DM, who will indicate his/her preference between $\mathbf{z}^{adj1}$ and the current solution $\mathbf{z}^{(1)}$.

  Suppose that the DM, reflecting an insecure attitude, hesitation or even ignorance regarding the question asked, says he/she does not know which of the two solutions he/she prefers.
- $x_6$ becomes basic and $\mathbf{z}^{adj2} = (12.5, 50, 25)$ is obtained.

  Suppose that, in this case, the answer of the DM is no, that is, the DM prefers solution $\mathbf{z}^{(1)}$.
- $x_3$ becomes basic and the solution $\mathbf{z}^{adj3} = (29, 3, 73)$ is obtained.

  Suppose that, in this case, the answer of the DM is yes, that is, he/she prefers solution $\mathbf{z}^{adj3}$.

Since the DM preferred one of the adjacent solutions, the algorithm goes to step 7 and $\mathbf{z}^a = (29, 3, 73)$ is candidate to be a final choice.

Step 7. Constraints are introduced on the weight space, based on the DM's answers to pairwise comparisons in step 3.

The first answer is he/she does not know, that is, the DM is not able to express his/her preference between $\mathbf{z}^{(1)}$ and $\mathbf{z}^{adj1}$; therefore, no constraint is introduced on the weight space.

The second answer is no, that is, the DM does not prefer $\mathbf{z}^{adj2}$ to $\mathbf{z}^{(1)}$; therefore, it leads to the introduction of the constraint:

$$\boldsymbol{\lambda}\left(\mathbf{z}^{(1)} - \mathbf{z}^{adj2}\right) \geq \varepsilon \Leftrightarrow [\lambda_1 \ \lambda_2 \ \lambda_3] \left( \begin{bmatrix} 18.33 \\ 15 \\ 71.67 \end{bmatrix} - \begin{bmatrix} 12.5 \\ 50 \\ 25 \end{bmatrix} \right) \geq \varepsilon,$$

where ε has a small positive value.

The third answer is yes, that is, the DM prefers $\mathbf{z}^{adj3}$ to $\mathbf{z}^{(1)}$; therefore, it leads to the constraint:

$$\boldsymbol{\lambda}\left(\mathbf{z}^{adj3} - \mathbf{z}^{(1)}\right) \geq \varepsilon \Leftrightarrow [\lambda_1 \ \lambda_2 \ \lambda_3] \left( \begin{bmatrix} 29 \\ 3 \\ 73 \end{bmatrix} - \begin{bmatrix} 18.33 \\ 15 \\ 71.67 \end{bmatrix} \right) \geq \varepsilon.$$

Figure 4.10 shows the constraints introduced on the weight space.

The reduced weight space $\Lambda^{(2)}$ is formed from $\Lambda^{(1)} = \Lambda$ by adding the constraints just introduced in this iteration.

Step 8. A "central" point $\boldsymbol{\lambda}^{(2)} \in \Lambda^{(2)}$ is determined in the reduced weight space. For this purpose, the following auxiliary problem that maximizes the smallest deviation to the frontier that defines $\Lambda^{(2)}$ is solved. A previous normalization of the constraint coefficients may be necessary to obtain a central point in the reduced weight space when these coefficients are in very different orders of magnitude (this is not required in this case).

Fig. 4.10 Constraints introduced on the weight space in step 7 of the first iteration

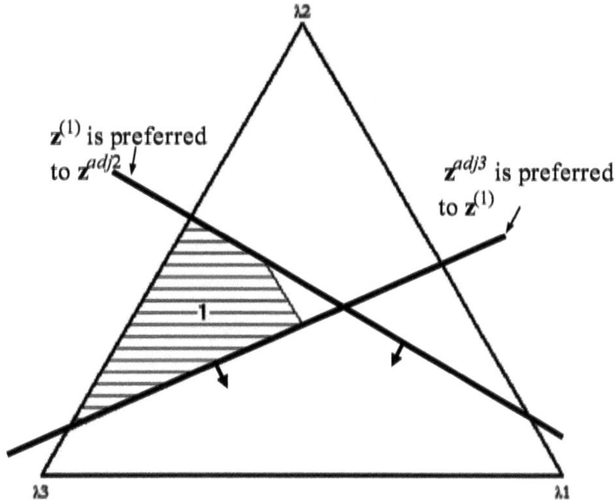


$$
\begin{aligned}
&\max\ \theta \\
&\text{s.t.} \\
&\qquad \lambda_1 + \lambda_2 + \lambda_3 = 1 \\
&\qquad \lambda_1 - \theta \geq \varepsilon \\
&\qquad \lambda_2 - \theta \geq \varepsilon \\
&\qquad \lambda_3 - \theta \geq \varepsilon \\
&\qquad 5.83\lambda_1 - 35\lambda_2 + 46.67\lambda_3 - \theta \geq \varepsilon \\
&\qquad 10.67\lambda_1 - 12\lambda_2 + 1.33\lambda_3 - \theta \geq \varepsilon \\
&\qquad \theta \geq 0
\end{aligned}
$$

where $\varepsilon$ is a very small positive scalar.

The solution to this problem is $\boldsymbol{\lambda}^{(2)} = (0.354, 0.323, 0.323)$, with $\theta = 0.323$.

Step 9. By using $\boldsymbol{\lambda}^{(2)}$, the following linear (weighted sum) programming problem is solved:

$$
\begin{aligned}
\max\quad & 0.354(3x_1 + x_2 + 2x_3 + x_4) \\
& +0.323(x_1 - x_2 + 2x_3 + 4x_4) \\
& +0.323(-x_1 + 5x_2 + x_3 + 2x_4) \\
\text{s.t.}\quad & 2x_1 + x_2 + 4x_3 + 3x_4 \leq 60 \\
& 3x_1 + 4x_2 + x_3 + 2x_4 \leq 60 \\
& x_1 + 2x_2 + 3x_3 + 4x_4 \leq 50 \\
& x_1, x_2, x_3, x_4 \geq 0
\end{aligned}
\qquad (P_{\lambda^2})
$$

The solution obtained is $\mathbf{x}^{(b)} = (14.5,\ 0,\ 2.5,\ 7)$, with $\mathbf{z}^{(b)} = (55.5,\ 47.5,\ 2)$.

Step 10. The solution $(\mathbf{x}^{(b)}, \mathbf{z}^{(b)})$ is presented to the DM who compares it with $\left(\mathbf{x}^{(a)}, \mathbf{z}^{(a)}\right) = \left(\mathbf{x}^{adj3}, \mathbf{z}^{adj3}\right)$, and expresses his/her preference. Suppose that the DM prefers $(\mathbf{x}^{(b)}, \mathbf{z}^{(b)})$.

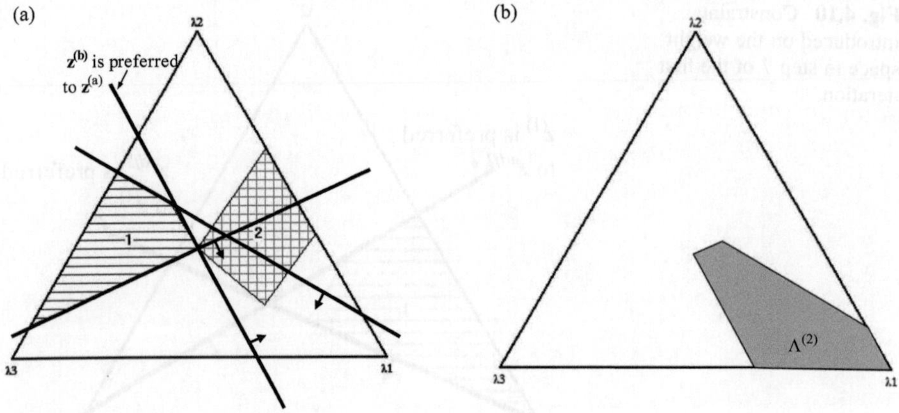


**Fig. 4.11** Solutions computed and constraints introduced on the weight space in the first iteration

A new constraint is then introduced on the reduced weight space $\Lambda^{(2)}$ (see Fig. 4.11a):

$$[\lambda_1 \ \lambda_2 \ \lambda_3]\left(\begin{bmatrix}55.5\\47.5\\2\end{bmatrix} - \begin{bmatrix}29\\3\\73\end{bmatrix}\right) \geq \varepsilon$$

Step 11. A new iteration, $h=2$, starts in which the new current solution is $\left(\mathbf{x}^{(2)}, \mathbf{z}^{(2)}\right) = \left(\mathbf{x}^{(b)}, \mathbf{z}^{(b)}\right)$.

The reduced weight space $\Lambda^{(2)}$ is defined by the original weight space $\Lambda$ with the constraints introduced in steps 7 and 10 of this iteration (Fig. 4.11b).

The method would continue for a second iteration by determining a central weight vector in the reduced weight space $\Lambda^{(2)}$ and obtaining a new efficient vertex through the optimization of the corresponding weighted-sum function.

The development of this second iteration can be found in the companion website.

## 4.4 TRIMAP

### 4.4.1 Method Presentation

TRIMAP is an interactive computational environment aimed at supporting the DM in the search for nondominated solutions. This method can be used in different ways, depending on the aims of the user. In this setting, TRIMAP is mainly a learning oriented computational environment more than a method in the usual sense.

The TRIMAP method, developed by Clímaco and Antunes (1987, 1989), combines a set of procedures allowing a free search based on the progressive and selective learning of the nondominated solution set. It combines the reduction of the feasible region with the reduction of the weight (parametric) space. The DM may specify lower bounds for the objective functions and/or impose constraints on the weight space. In each computation phase a weighted-sum of the objective functions is optimized.

The aim of the method is to help the DM progressively eliminating nondominated solutions that do not seem interesting to him/her. No convergence to any optimal solution of an implicit utility function is searched for. The interactive procedure only ends when the DM considers to have gathered 'sufficient knowledge' about the nondominated solution set, which enables him/her to make a final decision. TRIMAP combines three fundamental procedures: weight space decomposition, introduction of constraints in the objective function space and introduction of constraints on the weight space. The limitations introduced on the objective function values are automatically translated onto the weight space, which is used as a valuable means to gather and present the information to the DM. TRIMAP is dedicated to problems with three objective functions. Although this characteristic is a limitation, it allows the use of graphical means particularly useful in the dialogue with the DM. The main purpose is enabling a progressive and selective filling/exploitation of the weight space, which offers to the DM additional information about the nondominated region. The process stops when the DM feels comfortable to make a decision. In this way an exhaustive search of the solution set is avoided, namely of nondominated solution regions where the objective function values are very similar or considered uninteresting as a result of the information already gathered.

The reduction of the scope of the search is made through the imposition of limitations on the objective function values (a kind of information that is familiar to the DM, from the cognitive point of view). These constraints are automatically translated onto the weight space. The introduction of these additional limitations can also be used to obtain nondominated solutions that are not vertices of the feasible region. It is also possible to impose constraints directly on the weight space. This option is particularly interesting when TRIMAP is used as a teaching tool. The comparative analysis of the weight space and the objective function space during the interactive process enables the DM to decide on the interest of searching new nondominated solutions corresponding to regions of the weight space not yet explored.

The block diagram of the TRIMAP interactive environment is displayed in Fig. 4.12.

The use of TRIMAP combined with other interactive procedures is particularly suitable for a strategic search of nondominated solutions.

Initially, the nondominated solutions optimizing individually each objective function are computed, providing an overview of the range of variation of each objective function in the nondominated region. The nondominated solution that minimizes a weighted Chebyshev distance to the ideal solution can also be

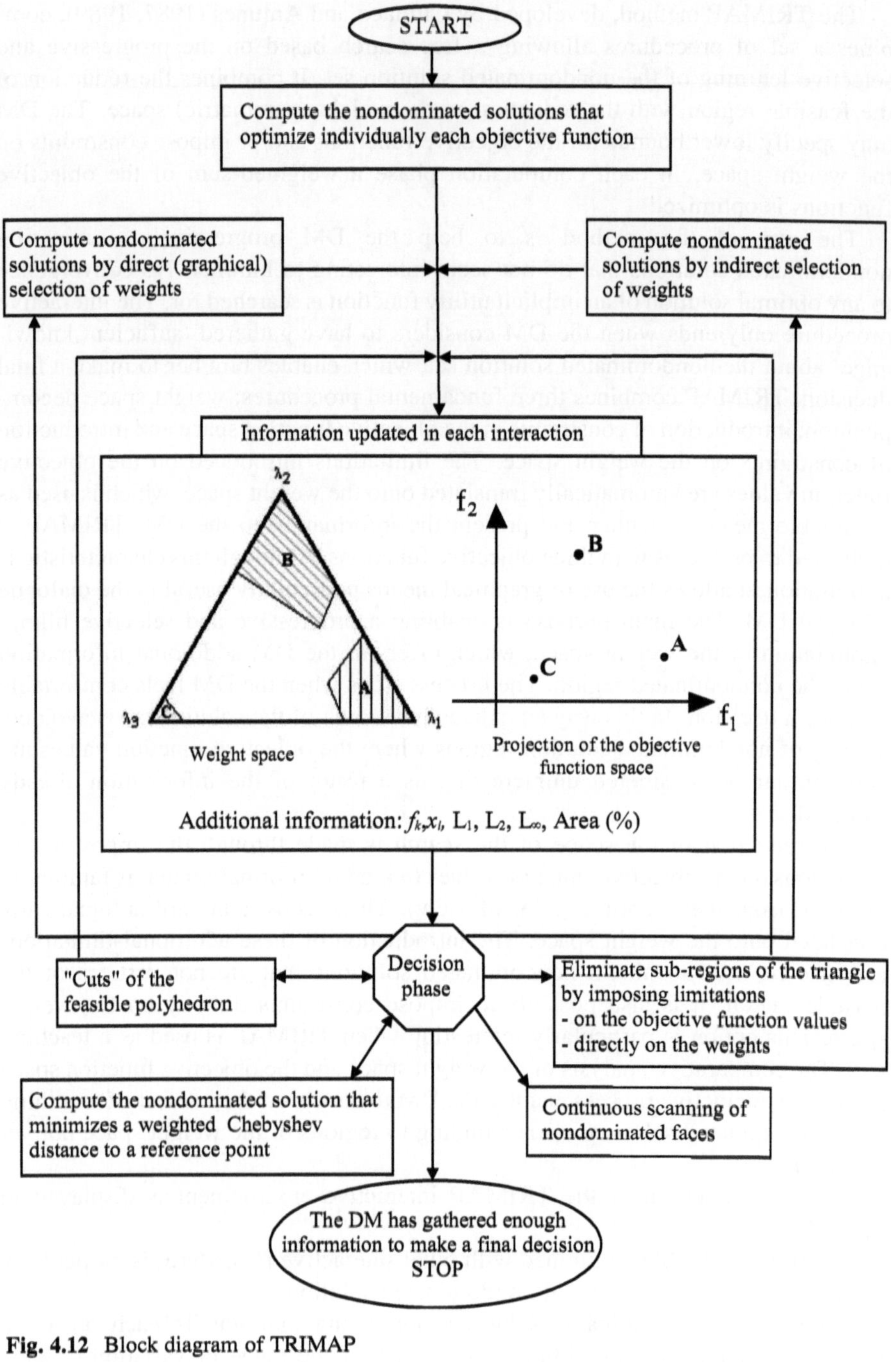


**Fig. 4.12** Block diagram of TRIMAP

computed, providing complementary information for the definition of the search direction for new nondominated solutions.

The **selection of weights** to compute new nondominated solutions can be made in two ways:

- Directly, when the DM chooses a new weight vector in the region of the triangle not yet filled, according to his/her options to continue the search;
- Indirectly, building a function whose gradient is normal to the plane defined by three nondominated solutions already computed and selected for this purpose by the DM. If some of the weights obtained using this process are not positive, a small perturbation in the gradient of the weighted objective function is made in order to assure this condition. This possibility of computing the weights is essentially the SIMOLP method of Reeves and Franz (1985).

The **introduction of additional limitations** on the objective function values, and the corresponding translation onto the weight space, allows the dialogue with the DM to be made in terms of the objective function values gathering the obtained information in the weight space. The introduction of an additional limitation, $f_k(\mathbf{x}) \geq L_k$ ($L_k \in \mathbb{R}$, $k \in \{1,2,3\}$), leads to the construction of the following *auxiliary* problem:

$$\begin{aligned} \max\ & z_k = f_k(\mathbf{x}) \\ \text{s.t. } & \mathbf{x} \in X_a \\ & X_a = \{\mathbf{x} \in X : f_k(\mathbf{x}) \leq L_k\} \end{aligned} \tag{4.2}$$

The maximization of $z_k = f_k(\mathbf{x})$ in $X_a$ leads to alternative optimal (basic) solutions (note that the gradient of the function is normal to the hyperplane that supports the auxiliary constraint $f_k(\mathbf{x}) \leq L_k$). Considering the objective functions of the original problem and the feasible region $X_a$, all efficient basic solutions that optimize (4.2) are computed. The corresponding indifference regions on the weight space are determined and graphically displayed. These are the indifference regions defined by $\boldsymbol{\lambda}\mathbf{W} \geq \mathbf{0}$, regarding each alternative efficient basis. The union of all these indifference regions determines the portion of the weight space where the additional limitation on the objective function value (imposed by the DM) is satisfied for the original problem. If the DM is only interested in nondominated solutions that satisfy $f_k(\mathbf{x}) \geq L_k$, then it is only necessary to restrict the search to this sub-region. If the DM wants to impose more than one limitation on the objective function values, then problem (4.2) is solved for each of them and the corresponding sub-regions on the weight space are filled with different patterns (or colours), thus allowing to clearly visualize the sub-regions of the weight space where there are intersections. The introduction of these additional limitations can also be used to obtain nondominated solutions that, in general, are not vertices of the original feasible region.

It is also possible to eliminate regions of the weight space by imposing direct limitations on the weights of the type $\frac{\lambda_i}{\lambda_j} \geq u_{ij}$, $i, j \in \{1,2,3\}$, $i \neq j$, $u_{ij} \in \mathbb{R}^+$, or $0 < u_L \leq \lambda_k \leq u_H < 1$ with $k \in \{1,2,3\}$.

The DM can also make **cuts** in the feasible region by establishing the values of one or two objective functions. Then, TRIMAP computes the nondominated solutions satisfying these additional constraints (which are not, in general, vertices of the feasible region). Notice that, when a cut $f_k(\mathbf{x}) = N_k$ $(N_k \in \mathbb{R})$ is made and nondominated vertices of the new problem (satisfying the additional constraint) are computed, it is possible to obtain dominated solutions regarding the original problem (for further details see Clímaco and Antunes (1987, 1989)).

TRIMAP also allows to **search for solutions in nondominated faces** between two previously computed nondominated points. This option is inspired by the *Pareto Race* method.

Moreover, the DM can identify a reference point in the objective function space and TRIMAP computes the nondominated solution that **minimizes a weighted Chebyshev distance** to that point.

In its original implementation (Clímaco and Antunes 1987, 1989), TRIMAP presents two main graphs. The first graph displays the weight (parametric) space, showing the indifference regions corresponding to the nondominated basic solutions already known. The second graph displays a projection of the objective function space, showing the nondominated solutions already computed. Other available graph is a *spider-web* diagram, displaying the differences between the objective function values of each solution and the corresponding components of a reference point (for instance, the ideal solution). Several complementary indicators are also available for each solution, namely: the $L_1$, $L_2$, and $L_\infty$ distances to the ideal solution and the area of the indifference regions (indicating the percentage occupied of the total area of the triangle).

In the next sections some potentialities of TRIMAP for teaching MOLP are emphasized and an illustrative example is presented to show how the main computation procedures used by TRIMAP operate.

### *4.4.2 Teaching Multiobjective Linear Programming Using TRIMAP*

In this section, the use of TRIMAP is illustrated through simple examples, in order to draw the readers' attention to some characteristics of MOLP problems.

Let us consider the following example:

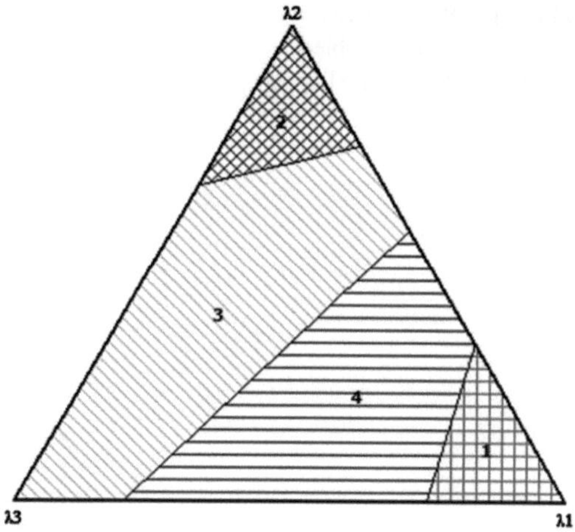


**Fig. 4.13** Weight space decomposition for problem (4.3)

$$\max_{\mathbf{x}\in X} \quad \mathbf{z}=f(\mathbf{x})=\begin{bmatrix} z_1=f_1(\mathbf{x}) \\ z_2=f_2(\mathbf{x}) \\ z_3=f_3(\mathbf{x}) \end{bmatrix} = \begin{bmatrix} -x_1+3x_2 \\ 2x_1-x_2 \\ 3x_1+2x_2 \end{bmatrix} \tag{4.3}$$

$$X=\{\mathbf{x}\in\mathbb{R}^2 : x_1\le 10,\ x_2\le 12,\ x_1+x_2\le 20,\ -x_1+x_2\le 7, \\ x_1-x_2\le 8,\ x_1\ge 0,\ x_2\ge 0\}.$$

The complete weight space decomposition into the indifference regions corresponding to the nondominated vertices of this problem can be computed using TRIMAP, as displayed in Fig. 4.13.

From this decomposition it can be concluded that it is possible to obtain the nondominated solution that optimizes $f_3(\mathbf{x})$, denoted by 3, by optimizing weighted sums of the objective functions that intuitively would not seem adequate to obtain that solution. For example, for $\lambda_1=0.36$, $\lambda_2=0.64$, $\lambda_3=0$, the optimum of $f_3(\mathbf{x})$ is obtained. This is a counter-intuitive result regarding the meaning of weights since the optimum of $f_3(\mathbf{x})$ is obtained with $\lambda_3=0$.

Note that, in this example, the weight space decomposition consists of indifference regions crossing the triangle from side to side ("stripy regions"). As the cone defined by the objective function gradients is flat, one of the functions can always be obtained as a linear combination of the other two. In these cases, the decomposition is always *striped* and counter-intuitive situations may occur, as the one indicated above.

Now, suppose that the first objective function is changed by multiplying $f_1(\mathbf{x})$ by 10, i.e. $z_1=f_1(\mathbf{x})=-10x_1+30x_2$, maintaining the remainder of (4.3). The resulting weight space decomposition is presented in Fig. 4.14.

The nondominated solutions of the previous problem are maintained, but the corresponding indifference regions are distorted as well as the relative percentages of the respective areas are noticeably changed.

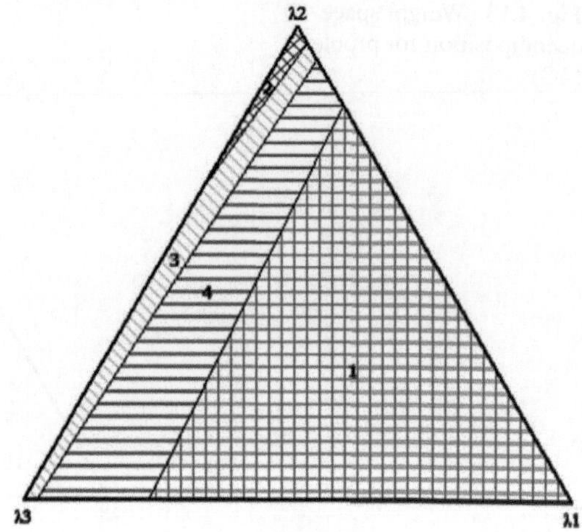


Fig. 4.14 Weight space decomposition of problem (4.3) by changing $f_1(\mathbf{x})$

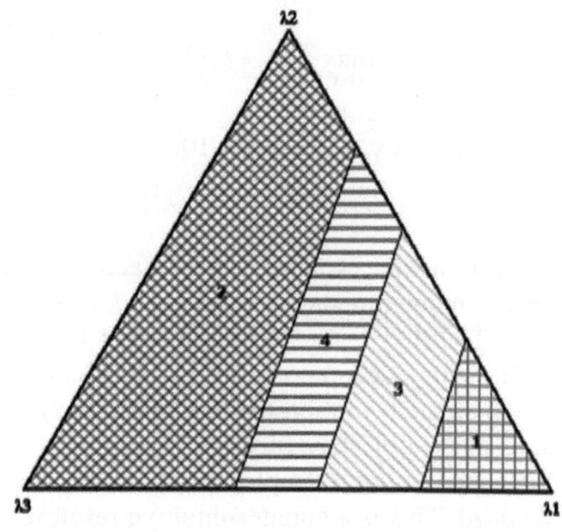


Fig. 4.15 Weight space decomposition of problem (4.3) modified by changing $f_3(\mathbf{x})$

Note that the change in the objective function $f_1(\mathbf{x})$ just alters the gradient magnitude. It can be concluded that, in order to give some practical meaning to the weight space decomposition and, therefore, to the set of weight vectors allowing the calculation of each nondominated vertex, it is advisable to make a previous normalization of the objective functions.

Finally, suppose that the initial problem (4.3) is maintained, except $f_3(\mathbf{x})$ that is changed to $z_3 = f_3(\mathbf{x}) = 3x_1 - 2x_2$. The new weight space decomposition is shown in Fig. 4.15.

The comparison of the weight space decomposition regarding problem (4.3), Fig. 4.13, with Fig. 4.15 allows concluding that, while in the original problem the optimal solutions of the three objective functions are distinct, now $f_2(\mathbf{x})$ and $f_3(\mathbf{x})$ have the same optimal solution.

### 4.4.3 Final Comments

TRIMAP allows a progressive and selective search of the nondominated solutions set in three-objective linear programming problems. In general, in the dialogue with the DM, he/she is asked to establish minimal satisfaction levels for the objective function values, by progressively delimiting the nondominated regions which are interesting to him/her, taking into account the existing knowledge about the nondominated set gathered throughout the interactive process. So, an exhaustive search of the nondominated solution set is avoided, thus saving computational effort, by making a progressive focus on the nondominated regions in which more interesting solutions for the DM are located.

The weight space and the optimization of weighted sums are essentially used for operational reasons. However, mediation of the dialogue with the DM by an analyst/facilitator is advisable due to the technical issues involved. Finally, we point out that in interactive decision processes based on TRIMAP there are no irrevocable decisions during the process, and it is not intended the convergence to the *best solution* of any implicit utility function.

### 4.4.4 Illustrative Example of the TRIMAP method

Consider the following linear programming problem with three objective functions:

$$\begin{aligned}
&\max\ z_1 = f_1(\mathbf{x}) = x_1\\
&\max\ z_2 = f_2(\mathbf{x}) = x_2\\
&\max\ z_3 = f_3(\mathbf{x}) = x_3\\
&\text{s.t.}\quad x_1 + x_2 + x_3 \leq 5\\
&\qquad\ \ x_1 + 3x_2 + x_3 \leq 9\\
&\qquad\ \ 3x_1 + 4x_2 \leq 16\\
&\qquad\ \ x_1, x_2, x_3 \geq 0
\end{aligned}$$

For a better understanding of the following procedures, the feasible region in the objective function space is presented in Fig. 4.16. Note that, in this case, this space coincides with the decision space.

In TRIMAP the nondominated solutions optimizing individually the 3 objective functions are automatically calculated and presented to the DM.

The optimization of weighted-sums of the objective functions constitutes the main computation procedure used in TRIMAP. This procedure entails solving the following linear programming problem:

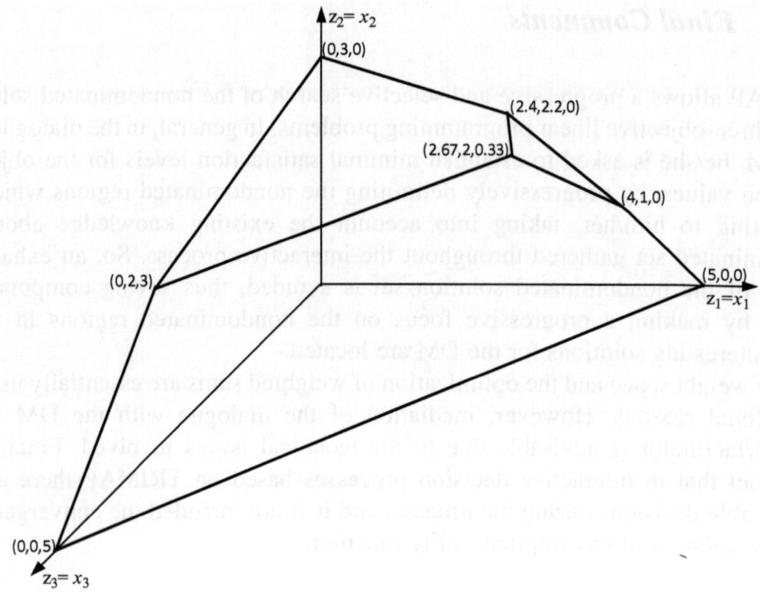


**Fig. 4.16** Decision space and objective function space

$$\begin{aligned} \max \quad & \sum_{k=1}^{3} \lambda_k \, f_k(\mathbf{x}) \\ \text{s.t.} \quad & \mathbf{x} \in X \end{aligned}$$

where $X$ is the feasible region of the problem and $\boldsymbol{\lambda} \in \Lambda$.

$\Lambda$ is the set of all the weight vectors and it is defined as:

$$\Lambda = \left\{ \boldsymbol{\lambda} \in \mathbb{R}^3 : \lambda_k > 0, \ k = 1, \ldots, 3; \ \sum_{k=1}^{3} \lambda_k = 1 \right\}$$

We start by computing the nondominated solution that optimizes $f_1(\mathbf{x})$ and the corresponding indifference region. In order to guarantee that the solution obtained is nondominated, the weights $\lambda_1 = 0.99$, $\lambda_2 = 0.005$ and $\lambda_3 = 0.005$ will be used instead of $\lambda_1 = 1$, $\lambda_2 = 0$ and $\lambda_3 = 0$ (since in this case just weakly nondominated solutions could be guaranteed).

The following problem is solved using the simplex method:

$$\begin{aligned} \max\ z &= 0.99x_1 + 0.005x_2 + 0.005x_3 \\ \text{s.t.}\quad & x_1 + x_2 + x_3 \leq 5 \\ & x_1 + 3x_2 + x_3 \leq 9 \\ & 3x_1 + 4x_2 \leq 16 \\ & x_1, x_2, x_3 \geq 0 \end{aligned}$$

In order to allow the subsequent computation of the indifference region, a reduced cost row associated with each objective function is added to the simplex tableau.

Let $x_4, x_5$ and $x_6$ be the slack variables.

The resolution by the simplex method is as follows:

| | **c** | 0.99 | 0.005 | 0.005 | 0 | 0 | 0 | |
|---|---|---|---|---|---|---|---|---|
| $(\mathbf{c_B})^T$ | $\mathbf{x_B}$ | $x_1$ | $x_2$ | $x_3$ | $x_4$ | $x_5$ | $x_6$ | |
| 0 | $x_4$ | 1 | 1 | 1 | 1 | 0 | 0 | 5 |
| 0 | $x_5$ | 1 | 3 | 1 | 0 | 1 | 0 | 9 |
| 0 | $x_6$ | 3 | 4 | 0 | 0 | 0 | 1 | 16 |
| $z_j^1 - c_j^1$ | | −1 | 0 | 0 | 0 | 0 | 0 | 0 |
| $z_j^2 - c_j^2$ | | 0 | −1 | 0 | 0 | 0 | 0 | 0 |
| $z_j^3 - c_j^3$ | | 0 | 0 | −1 | 0 | 0 | 0 | 0 |
| $z_j^\lambda - c_j^\lambda$ | | −0.99 | −0.005 | −0.05 | 0 | 0 | 0 | |

| | **c** | 0.99 | 0.005 | 0.005 | 0 | 0 | 0 | |
|---|---|---|---|---|---|---|---|---|
| $(\mathbf{c_B})^T$ | $\mathbf{x_B}$ | $x_1$ | $x_2$ | $x_3$ | $x_4$ | $x_5$ | $x_6$ | |
| 0.99 | $x_1$ | 1 | 1 | 1 | 1 | 0 | 0 | 5 |
| 0 | $x_5$ | 0 | 2 | 0 | −1 | 1 | 0 | 4 |
| 0 | $x_6$ | 0 | 1 | −3 | −3 | 0 | 1 | 1 |
| $z_j^1 - c_j^1$ | | 0 | 1 | 1 | 1 | 0 | 0 | 5 |
| $z_j^2 - c_j^2$ | | 0 | −1 | 0 | 0 | 0 | 0 | 0 |
| $z_j^3 - c_j^3$ | | 0 | 0 | −1 | 0 | 0 | 0 | 0 |
| $z_j^\lambda - c_j^\lambda$ | | 0 | 0.985 | 0.985 | 0.99 | 0 | 0 | |

The optimal solution to the weighted-sum problem is found. This is the nondominated solution optimizing $f_1(\mathbf{x})$ in the multiobjective problem:

$$x_1 = 5; x_2 = 0; x_3 = 0;$$
$$z_1 = 5; z_2 = 0; z_3 = 0;$$

Note that the values in the row $z_j^\lambda - c_j^\lambda$ are equal to $\lambda_1\left(z_j^1 - c_j^1\right) + \lambda_2\left(z_j^2 - c_j^2\right)$ $+\lambda_3\left(z_j^3 - c_j^3\right)$ for the specified weight vector (in this case, $(\lambda_1, \lambda_2, \lambda_3) =$ (0.99, 0.005, 0.005) ). All the weight vectors that lead to this nondominated

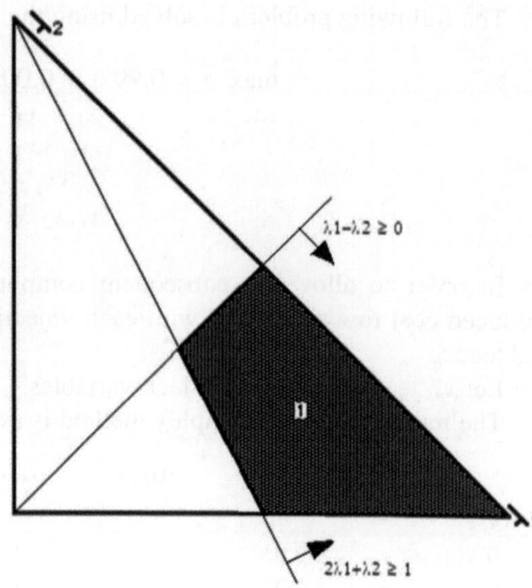


**Fig. 4.17** Indifference region of the solution that optimizes $f_1(\mathbf{x})$

solution (indifference region) can be determined by verifying in what conditions all $z_j^\lambda - c_j^\lambda \geq 0$ are obtained (optimality condition in the simplex method), that is: $\lambda_1\left(z_j^1 - c_j^1\right) + \lambda_2\left(z_j^2 - c_j^2\right) + \lambda_3\left(z_j^3 - c_j^3\right) \geq 0$. Thus, for the current solution, the indifference region is given by:

$$\Lambda = \left\{\boldsymbol{\lambda} \in \mathbb{R}^3 : \lambda_k > 0, \quad k = 1,..,3; \quad \sum_{k=1}^{3} \lambda_k = 1\right\} \quad \text{and}$$

$\lambda_1 - \lambda_2 \geq 0$ (from the column of the reduced cost matrix corresponding to $x_2$),
$\lambda_1 - \lambda_3 \geq 0$ (from the column of the reduced cost matrix corresponding to $x_3$), and
$\lambda_1 \geq 0$ (from the column of the reduced cost matrix corresponding to $x_4$).

In order to represent this region on the projection of the weight space $(\lambda_1, \lambda_2)$, the variable $\lambda_3$ is substituted in $\lambda_1 - \lambda_3 \geq 0$, making $\lambda_3 = 1 - \lambda_1 - \lambda_2$ (since $\sum_{k=1}^{3} \lambda_k = 1$). Then, this inequality is equivalent to $2\lambda_1 + \lambda_2 \geq 1$.

The indifference region corresponding to the solution that optimizes $f_1(\mathbf{x})$ is presented in Fig. 4.17.[2]

Using a similar process, TRIMAP also computes the nondominated solutions that optimize $f_2(\mathbf{x})$ and $f_3(\mathbf{x})$. These solutions are presented to the DM by using two

[2] Most figures in this illustrative example are screen copies of the TRIMAP package for Macintosh (Clímaco and Antunes 1989).

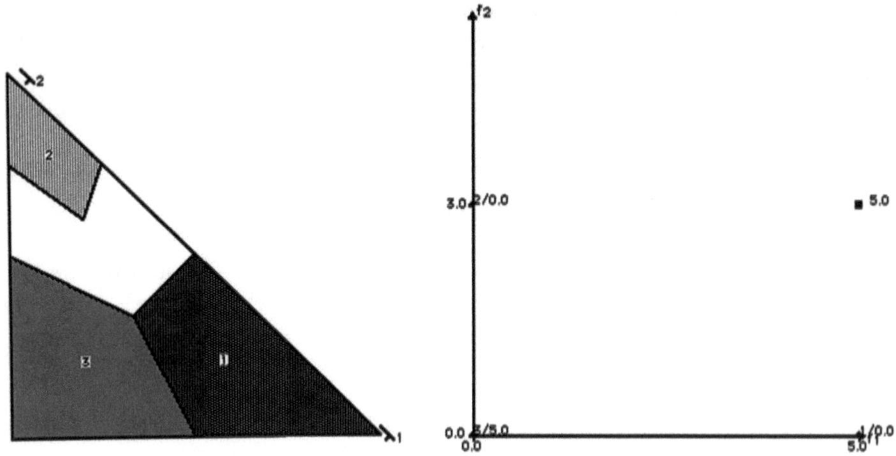


**Fig. 4.18** Solutions optimizing individually each objective function

graphs: the weight space (or a projection of it) and the two-dimensional projection of the objective function space. Figure 4.18 shows the projection of the weight space onto $(\lambda_1, \lambda_2)$ and the two-dimensional projection of the objective space onto $(f_1, f_2)$. In this graph, the value $f_3$ is shown near the solution index and the ideal solution is represented by a small black square. From Fig. 4.18 it can be concluded that there is a nondominated edge connecting solutions (vertices) 1 and 3, because the corresponding indifference regions are contiguous, that is, they have a common edge.

Solution 1: $\mathbf{x}^1 = \mathbf{z}^1 = (5, 0, 0)$;
Solution 2: $\mathbf{x}^2 = \mathbf{z}^2 = (0, 3, 0)$;
Solution 3: $\mathbf{x}^3 = \mathbf{z}^3 = (0, 0, 5)$;

The existence of a single indifference region including each vertex of the triangle, Fig. 4.18, indicates that no alternative nondominated optima exist for each objective function. We will see below an example where alternative nondominated optima exist.

Moreover, this problem is non-degenerate. In this case, all nondominated vertices will be known only when the triangle is fully filled because there is a one-to-one correspondence between indifference regions and nondominated vertices. Whenever degenerate solutions exist, more than one basic solution may correspond to a given vertex and, therefore, the decomposition of the weight space becomes more complicated. In fact, in many degenerate problems it is possible to obtain all the nondominated vertices without completely filling the triangle. The basic solutions corresponding to the same nondominated vertex may correspond to different indifference regions, possibly partially overlapping. This situation does not inhibit but complicates the use of TRIMAP. An example of a multiobjective problem with degenerate solutions is shown in Chap. 5.

Other nondominated vertex solutions can be computed by direct or indirect selection made by the DM of other weight vectors belonging to the region of the triangle not yet filled.

Suppose that the DM wants to know the nondominated solution that is obtained by optimizing a weighted function whose gradient is normal to the plane defined by the three solutions already computed (indirect selection of weights). The computation of these weights is performed as follows:

Let $\mathbf{v}^1$ and $\mathbf{v}^2$ be two vectors of the objective function space defined by $\mathbf{v}^1 = \mathbf{z}^2 - \mathbf{z}^1 = (-5, 3, 0)$ and $\mathbf{v}^2 = \mathbf{z}^3 - \mathbf{z}^1 = (-5, 0, 5)$. These two vectors define a plane parallel to the one defined by the points $\mathbf{z}^1$, $\mathbf{z}^2$ and $\mathbf{z}^3$. Then, the vector $\mathbf{v} = (v_1, v_2, v_3)$, where $v_1 = \begin{vmatrix} 3 & 0 \\ 0 & 5 \end{vmatrix} = 15$, $v_2 = (-1)\begin{vmatrix} -5 & 0 \\ -5 & 5 \end{vmatrix} = 25$ and $v_3 = \begin{vmatrix} -5 & 3 \\ -5 & 0 \end{vmatrix} = 15$, is normal to the plane defined by $\mathbf{v}^1$ and $\mathbf{v}^2$. Since the weighted-sum function is $\lambda_1\ f_1(\mathbf{x}) + \lambda_2\ f_2(\mathbf{x}) + \lambda_3\ f_3(\mathbf{x})$, the corresponding gradient in the objective space is given by $(\lambda_1, \lambda_2, \lambda_3)$. Hence, $\boldsymbol{\lambda}$ will be equal to the normalized vector $\mathbf{v}$, that is (0.273, 0.455, 0.273).

The following weighted-sum problem is solved:

$$\begin{aligned} \max\ z = {} & 0.273x_1 + 0.455x_2 + 0.273x_3 \\ \text{s.t.}\quad & x_1 + x_2 + x_3 \le 5 \\ & x_1 + 3x_2 + x_3 \le 9 \\ & 3x_1 + 4x_2 \le 16 \\ & x_1, x_2, x_3 \ge 0 \end{aligned}$$

The solution of this problem is the nondominated solution 4, $\mathbf{x}^4 = \mathbf{z}^4 = (2.67, 2, 0.33)$, whose indifference region is presented in Fig. 4.19.

Figure 4.19 allows concluding that solutions 1, 3 and 4 belong to a nondominated face (because there is a point in interior of the weight space that is common to the indifference regions of these 3 solutions). Furthermore, solutions 2 and 4 also belong to another nondominated face.

The imposition of bounds on the objective function values and its translation into the weight space is a TRIMAP procedure particularly useful to the DM. The imposition of additional bounds allows reducing the feasible region and, therefore, reducing the scope of the search for new nondominated solutions.

When the DM imposes one or more bounds on the objective function values, that information is translated into the weight space: the area corresponding to the weight vectors that lead to solutions that satisfy this (these) limitation(s) is presented. Suppose that the DM decides to impose the lower bound $f_3(\mathbf{x}) \ge 2$. In order to compute the region of the weight space where this constraint is satisfied, the following *auxiliary* problem is solved:

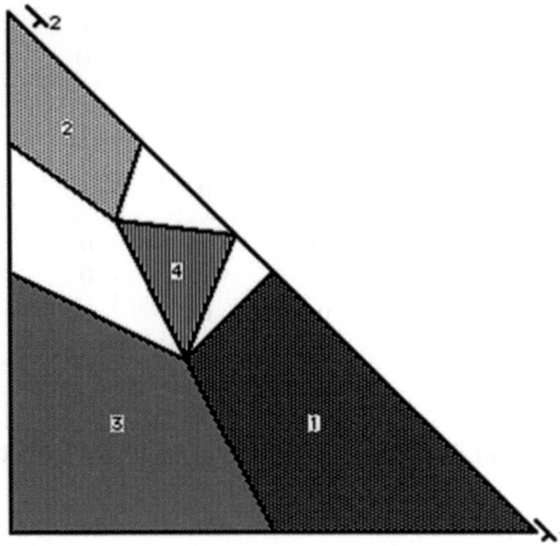


**Fig. 4.19** Weight space decomposition after computing solution 4

$$\begin{aligned} &\max \ f_3(\mathbf{x}) \\ &\text{s.t.} \ \ \mathbf{x} \in X \\ &\quad\ f_3(\mathbf{x}) \leq 2 \end{aligned}$$

Note that the constraint introduced in the *auxiliary* problem, $f_3(\mathbf{x}) \leq 2$, is opposed to the limitation imposed by the DM, i.e. $f_3(\mathbf{x}) \geq 2$, thus leading to alternative optimal solutions. All these alternative optima are computed as well as the corresponding indifference regions of the *multiobjective auxiliary* problem:

$$\begin{aligned} &\max \ f_1(\mathbf{x}) \\ &\max \ f_2(\mathbf{x}) \\ &\max \ f_3(\mathbf{x}) \\ &\text{s.t.} \quad \mathbf{x} \in X \\ &\quad\ f_3(\mathbf{x}) \leq 2 \end{aligned}$$

The union of these indifference regions defines the region of the weight space where $f_3(\mathbf{x}) \geq 2$.

The *auxiliary* problem associated with the additional constraint is:

$$\begin{aligned} \max \ z_3 = f_3(\mathbf{x}) &= x_3 \\ \text{s.t.} \qquad x_1 + x_2 + x_3 &\leq 5 \\ x_1 + 3x_2 + x_3 &\leq 9 \\ 3x_1 + 4x_2 &\leq 16 \\ x_3 &\leq 2 \\ x_1, x_2, x_3 &\geq 0 \end{aligned}$$

The resolution of this problem using the simplex method is as follows: ($x_4$, $x_5$, $x_6$ and $x_7$ are slack variables)

| $(\mathbf{c}_B)^T$ | $\mathbf{c}$ / $\mathbf{x}_B$ | 0 / $x_1$ | 0 / $x_2$ | 1 / $x_3$ | 0 / $x_4$ | 0 / $x_5$ | 0 / $x_6$ | 0 / $x_7$ | |
|---|---|---|---|---|---|---|---|---|---|
| 0 | $x_4$ | 1 | 1 | 1 | 1 | 0 | 0 | 0 | 5 |
| 0 | $x_5$ | 1 | 3 | 1 | 0 | 1 | 0 | 0 | 9 |
| 0 | $x_6$ | 3 | 4 | 0 | 0 | 0 | 1 | 0 | 16 |
| 0 | $x_7$ | 0 | 0 | **1** | 0 | 0 | 0 | 1 | 2 |
| $z_j^1 - c_j^1$ | | −1 | 0 | 0 | 0 | 0 | 0 | 0 | 0 |
| $z_j^2 - c_j^2$ | | 0 | −1 | 0 | 0 | 0 | 0 | 0 | 0 |
| $z_j^3 - c_j^3$ | | 0 | 0 | −1 | 0 | 0 | 0 | 0 | 0 |
| $z_j - c_j$ | | 0 | 0 | **−1** | 0 | 0 | 0 | 0 | |

Second iteration ($x_3$ becomes basic and $x_7$ becomes nonbasic):

| $(\mathbf{c}_B)^T$ | $\mathbf{c}$ / $\mathbf{x}_B$ | 0 / $x_1$ | 0 / $x_2$ | 1 / $x_3$ | 0 / $x_4$ | 0 / $x_5$ | 0 / $x_6$ | 0 / $x_7$ | |
|---|---|---|---|---|---|---|---|---|---|
| 0 | $x_4$ | 1 | 1 | 0 | 1 | 0 | 0 | −1 | 3 |
| 0 | $x_5$ | 1 | 3 | 0 | 0 | 1 | 0 | −1 | 7 |
| 0 | $x_6$ | 3 | 4 | 0 | 0 | 0 | 1 | 0 | 16 |
| 1 | $x_3$ | 0 | 0 | 1 | 0 | 0 | 0 | 1 | 2 |
| $z_j^1 - c_j^1$ | | −1 | 0 | 0 | 0 | 0 | 0 | 0 | 0 |
| $z_j^2 - c_j^2$ | | 0 | −1 | 0 | 0 | 0 | 0 | 0 | 0 |
| $z_j^3 - c_j^3$ | | 0 | 0 | 0 | 0 | 0 | 0 | 1 | 2 |
| $z_j - c_j$ | | **0** | **0** | 0 | 0 | 0 | 0 | 1 | |

This is one of the optimal solutions to the *auxiliary* problem, but there are alternative optima, as it can be verified from the corresponding optimal tableau, since there are 2 values equal to 0 in the row $z_j - c_j$ corresponding to the nonbasic variables $x_1$ and $x_2$. The solution obtained ($x_1 = 0, x_2 = 0, x_3 = 2$) is not an efficient solution to the *multiobjective auxiliary* problem. The computation of the corresponding indifference region using the conditions obtained from the multiobjective simplex tableau enables to conclude that there is no weight vector that satisfies all these conditions. This can be verified as follows:

$$\Lambda = \left\{ \boldsymbol{\lambda} \in \mathbb{R}^3 : \lambda_k > 0, \quad k = 1,..,3; \quad \sum_{k=1}^{3} \lambda_k = 1 \right\} \quad \text{and}$$

$-\lambda_1 \geq 0$ (from the column of the reduced cost matrix corresponding to $x_1$),
$-\lambda_2 \geq 0$ (from the column of the reduced cost matrix corresponding to $x_2$), and
$\lambda_3 \geq 0$ (from the column of the reduced cost matrix corresponding to $x_7$)

The intersection of these conditions defines an empty set (because each weight is strictly positive).

Then it is necessary to compute all the alternative optima of the previous solution (for a more detailed study, see the algorithm for computing all alternative optimal solutions to an LP proposed by Steuer (1986, chapter 4)).

There are two pivot options:
1. making $x_1$ basic and $x_4$ nonbasic;
2. making $x_2$ basic and $x_5$ nonbasic.

Let us start by the first option:

| $(\mathbf{c}_B)^T$ | $\mathbf{c}$ / $\mathbf{x}_B$ | 0 / $x_1$ | 0 / $x_2$ | 1 / $x_3$ | 0 / $x_4$ | 0 / $x_5$ | 0 / $x_6$ | 0 / $x_7$ | |
|---|---|---|---|---|---|---|---|---|---|
| 0 | $x_1$ | 1 | 1 | 0 | 1 | 0 | 0 | -1 | 3 |
| 0 | $x_5$ | 0 | 2 | 0 | −1 | 1 | 0 | 0 | 4 |
| 0 | $x_6$ | 0 | 1 | 0 | −3 | 0 | 1 | 3 | 7 |
| 1 | $x_3$ | 0 | 0 | 1 | 0 | 0 | 0 | 1 | 2 |
| $z_j^1 - c_j^1$ | | 0 | 1 | 0 | 1 | 0 | 0 | −1 | 3 |
| $z_j^2 - c_j^2$ | | 0 | -1 | 0 | 0 | 0 | 0 | 0 | 0 |
| $z_j^3 - c_j^3$ | | 0 | 0 | 0 | 0 | 0 | 0 | 1 | 2 |
| $z_j - c_j$ | | 0 | **0** | 0 | 0 | 0 | 0 | 1 | |

Let this solution be designated by A, where $x_1 = 3$, $x_2 = 0$, $x_3 = 2$ and $z_1 = 3$, $z_2 = 0$, $z_3 = 2$.

The computation of the indifference region proceeds as follows:

$$\Lambda = \left\{ \boldsymbol{\lambda} \in \mathbb{R}^3 : \lambda_k > 0, \quad k = 1, .., 3; \quad \sum_{k=1}^{3} \lambda_k = 1 \right\} \quad \text{and}$$

$\lambda_1 - \lambda_2 \geq 0$ (from the column of the reduced cost matrix corresponding to $x_2$)
$\lambda_1 \geq 0$ (from the column of the reduced cost matrix corresponding to $x_4$)
$-\lambda_1 + \lambda_3 \geq 0$ (from the column of the reduced cost matrix corresponding to $x_7$)

From $-\lambda_1 + \lambda_3 \geq 0 \Leftrightarrow 2\lambda_1 + \lambda_2 \leq 1$ the indifference region corresponding to solution A can be graphically presented, as it is shown in Fig. 4.20. Solution A has a non-empty indifference region, guaranteeing that it is an efficient solution to the *multiobjective auxiliary* problem.

Let us continue computing the alternative optimal solutions to the *auxiliary* problem, considering now the second pivot option, from the first optimal tableau ($x_2$ becomes basic and $x_5$ nonbasic):

| $(\mathbf{c}_B)^T$ | $\mathbf{c}$ / $\mathbf{x}_B$ | 0 / $x_1$ | 0 / $x_2$ | 1 / $x_3$ | 0 / $x_4$ | 0 / $x_5$ | 0 / $x_6$ | 0 / $x_7$ | |
|---|---|---|---|---|---|---|---|---|---|
| 0 | $x_4$ | $2/3$ | 0 | 0 | 1 | $-1/3$ | 0 | $-2/3$ | $2/3$ |
| 0 | $x_2$ | $1/3$ | 1 | 0 | 0 | $1/3$ | 0 | $-1/3$ | $7/3$ |
| 0 | $x_6$ | $5/3$ | 0 | 0 | 0 | $-4/3$ | 1 | $4/3$ | $20/3$ |
| 1 | $x_3$ | 0 | 0 | 1 | 0 | 0 | 0 | 1 | 2 |
| $z_j^1 - c_j^1$ | | −1 | 0 | 0 | 0 | 0 | 0 | 0 | 0 |
| $z_j^2 - c_j^2$ | | $1/3$ | 0 | 0 | 0 | $1/3$ | 0 | $-1/3$ | $7/3$ |
| $z_j^3 - c_j^3$ | | 0 | 0 | 0 | 0 | 0 | 0 | 1 | 2 |
| $z_j - c_j$ | | 0 | 0 | 0 | 0 | 0 | 0 | 1 | |

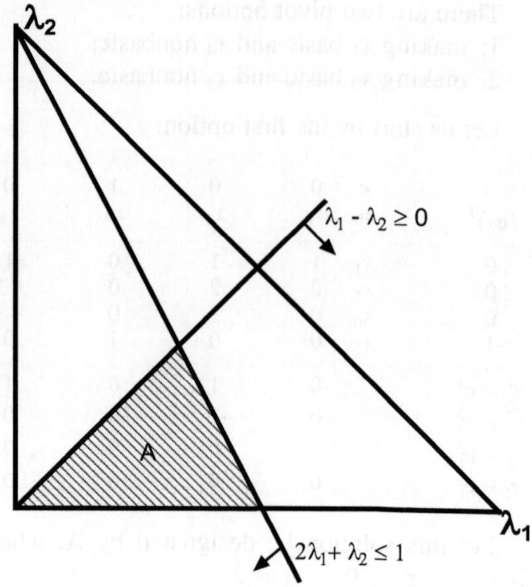


**Fig. 4.20** Solution A is a nondominated solution to the *multiobjective auxiliary* problem

Let us designate this solution by B, where $x_1 = 0,\ x_2 = \frac{7}{3},\quad x_3 = 2$ and $z_1 = 0,\ z_2 = \frac{7}{3},\ z_3 = 2$. The corresponding indifference region is defined by:

$$\Lambda^{\mathrm{B}} = \left\{ \boldsymbol{\lambda} \in \mathbb{R}^3 : \lambda_k > 0,\quad k = 1,..,3;\quad \sum_{k=1}^{3} \lambda_k = 1,\quad - \lambda_1 + \frac{1}{3}\lambda_2 \geq 0, \right.$$

$$\left. \frac{1}{3}\lambda_2 \geq 0,\quad - \frac{1}{3}\lambda_2 + \lambda_3 \geq 0 \right\}$$

This region is displayed in Fig. 4.21.

The computation of alternative optima can still be continued because the possibilities of obtaining different alternative optimal bases have not yet been completely exploited. From the optimal tableau of solution A another alternative optimal solution is computed, by making $x_2$ basic and $x_5$ nonbasic:

| $(\mathbf{c}_B)^T$ | $\mathbf{c}$ / $\mathbf{x}_B$ | 0 / $x_1$ | 0 / $x_2$ | 1 / $x_3$ | 0 / $x_4$ | 0 / $x_5$ | 0 / $x_6$ | 0 / $x_7$ | |
|---|---|---|---|---|---|---|---|---|---|
| 0 | $x_1$ | 1 | 0 | 0 | $^3/_2$ | $-^1/_2$ | 0 | -1 | 1 |
| 0 | $x_2$ | 0 | 1 | 0 | $-^1/_2$ | $^1/_2$ | 0 | 0 | 2 |
| 0 | $x_6$ | 0 | 0 | 0 | $-^5/_2$ | $-^1/_2$ | 1 | 3 | 5 |
| 1 | $x_3$ | 0 | 0 | 1 | 0 | 0 | 0 | 1 | 2 |
| $z_j^1 - c_j^1$ | | 0 | 0 | 0 | $^3/_2$ | $-^1/_2$ | 0 | -1 | 1 |
| $z_j^2 - c_j^2$ | | 0 | 0 | 0 | $-^1/_2$ | $^1/_2$ | 0 | 0 | 2 |
| $z_j^3 - c_j^3$ | | 0 | 0 | 0 | 0 | 0 | 0 | 1 | 2 |
| $z_j - c_j$ | | 0 | 0 | 0 | 0 | 0 | 0 | 1 | |

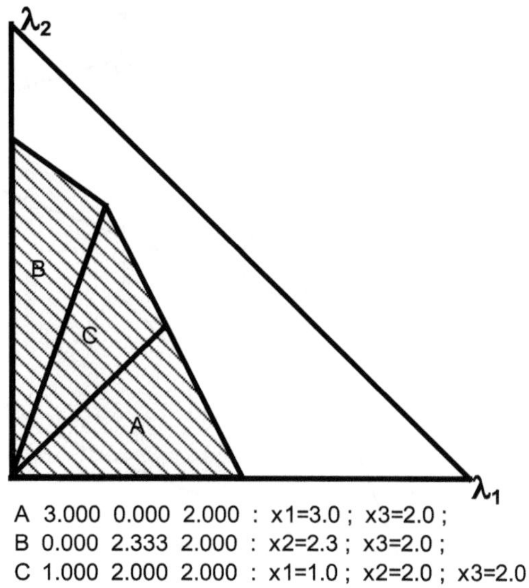


**Fig. 4.21** Indifference regions of the alternative optima of $f_3(\mathbf{x})$, for the *multiobjective auxiliary* problem

Let us designate this solution by C, where $x_1 = 1; \quad x_2 = 2; \quad x_3 = 2$ and $z_1 = 1; \quad z_2 = 2; \quad z_3 = 2$. The indifference region is defined by:

$$\Lambda^{C} = \Big\{\boldsymbol{\lambda} \in \mathbb{R}^3 : \lambda_k > 0, \quad k = 1,..,3; \quad \sum_{k=1}^{3} \lambda_k = 1, \ \tfrac{3}{2}\lambda_1 - \tfrac{1}{2}\lambda_2 \geq 0,$$
$$-\tfrac{1}{2}\lambda_1 + \tfrac{1}{2}\lambda_2 \geq 0, \quad -\lambda_1 + \lambda_3 \geq 0\Big\}$$

This region is also displayed in Fig. 4.21.

Since all the possibilities of obtaining alternative optima have been considered, TRIMAP presents the weight space region where the limitation $f_3(\mathbf{x}) \geq 2$ is satisfied (Fig. 4.21). This region is composed by the union of the indifference regions corresponding to the alternative optimal solutions to the *auxiliary* problem, which are nondominated to the original multiobjective problem with the auxiliary constraint ($f_3(\mathbf{x}) \leq 2$), that is, the *multiobjective auxiliary* problem. Note that, in certain cases, nondominated solutions may exist for the *multiobjective auxiliary* problem that are dominated for the original problem.

The visual inspection of the sub-regions A, B and C in Fig. 4.21 confirms the existence of alternative optima for $f_3(\mathbf{x})$. Note that all these sub-regions include the vertex of the triangle where $\lambda_1 = 0, \lambda_2 = 0, \lambda_3 = 1$.

In Fig. 4.22 the objective space is presented, indicating with a dashed pattern the non-dominated frontier resulting from the imposition of the constraint $f_3(\mathbf{x}) \geq 2$.

Note that the first optimal solution obtained by solving the *auxiliary* problem, which is not nondominated to the original problem, is indicated by Y in Fig. 4.22. Solutions A, B and C are nondominated basic solutions for the multiobjective problem with $f_3(\mathbf{x}) \geq 2$, although they are nonbasic to the original problem.

Suppose that the DM wants to search for a nondominated solution satisfying the additional constraint $f_3(\mathbf{x}) \geq 2$. For this purpose, he/she selects a weight vector

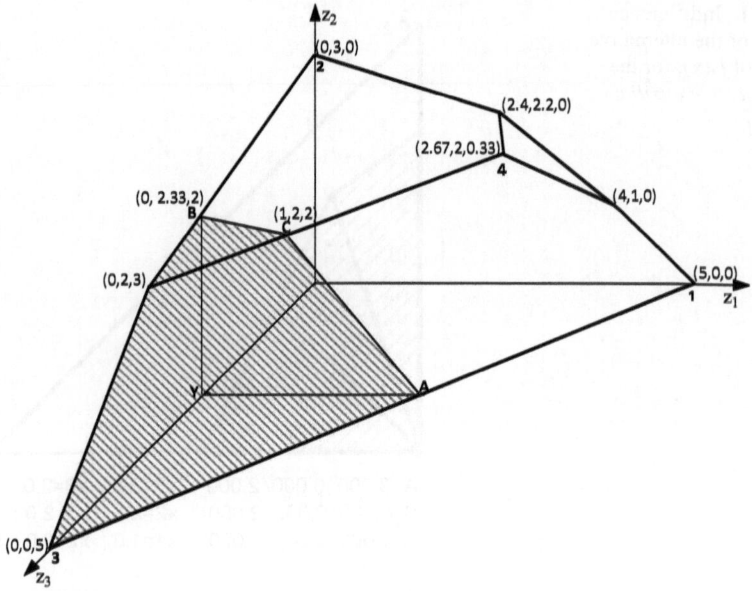


**Fig. 4.22** Nondominated solutions which satisfy $f_3(\mathbf{x}) \geq 2$

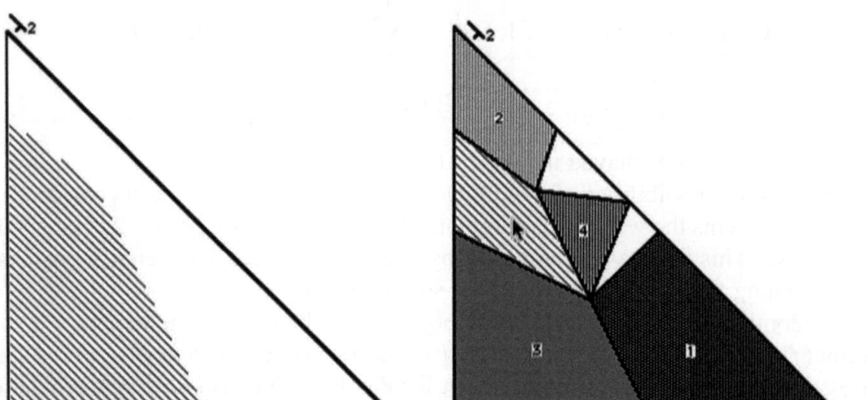


**Fig. 4.23** Selecting a weight vector to compute a new solution satisfying $f_3(\mathbf{x}) \geq 2$

within the dashed area of the triangle (union of the indifference regions in Fig. 4.21)—see Fig. 4.23.

Suppose the choice is the weight vector (0.129, 0.550, 0.321)—Fig. 4.23. The problem to be solved is:

$$
\begin{aligned}
\max \quad & z = 0.129f_1(\mathbf{x}) + 0.550f_2(\mathbf{x}) + 0.321f_3(\mathbf{x}) = 0.129x_1 + 0.550x_2 + 0.321x_3 \\
\text{s.t.} \quad & x_1 + x_2 + x_3 \leq 5 \\
& x_1 + 3x_2 + x_3 \leq 9 \\
& 3x_1 + 4x_2 \leq 16 \\
& x_1, x_2, x_3 \geq 0
\end{aligned}
$$

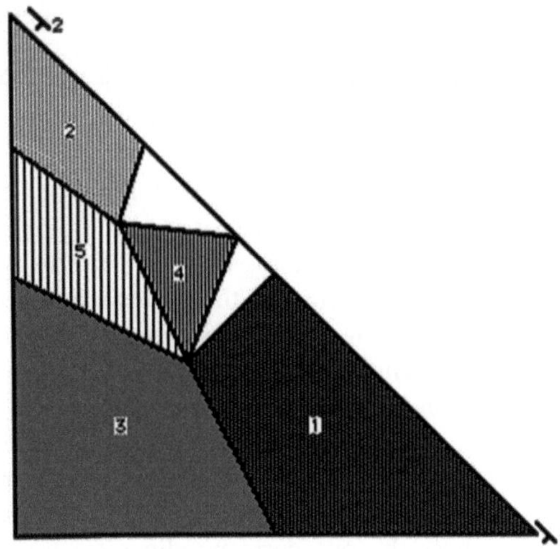


**Fig. 4.24** Weight space decomposition after computing solution 5

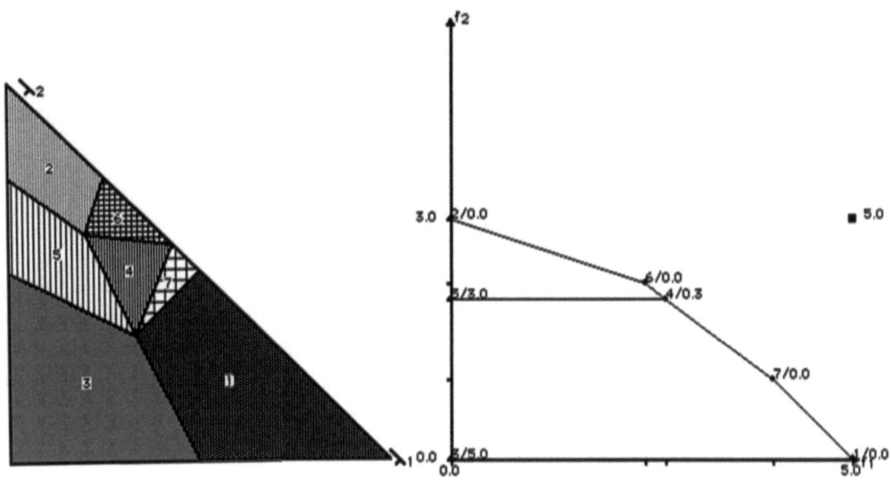


**Fig. 4.25** All nondominated vertices are known

A new nondominated solution is obtained, solution 5, where $\mathbf{x}^5 = \mathbf{z}^5 = (0,\ 2,\ 3)$—Fig. 4.24.

If the DM wants to know all the other nondominated solutions that completely fill the triangle, then all nondominated basic solutions to the original multiobjective problem are calculated. The nondominated faces and edges can be identified through the analysis of the graphs presented by TRIMAP.

Figure 4.25 shows the complete decomposition of the weight space. As the problem under study is non-degenerate the entire set of nondominated vertices and therefore the entire set of nondominated solutions can be obtained from the information contained in the two graphs in Fig. 4.25.

Note that, in general, the purpose of TRIMAP is not to generate the entire set of nondominated solutions.

The selective part of the progressive and selective search developed by TRIMAP can be materialized through the introduction of constraints on the objective function values, which are automatically translated onto the weight space. However, the visual inspection of the two diagrams of Fig. 4.25 throughout the search can also advise the analyst on the possible relinquishing of certain regions of the triangle that have not been explored yet. In fact, if, for instance, during the search process, solutions 2, 3 and 4 are already known and the corresponding objective function values are considered very similar, it could not be worthwhile to search the hatched weight space region where the cursor is located in Fig. 4.23. In fact, as the feasible region is a convex polyhedron, solutions that could be found in that region would not add further information. The elimination of the regions of the triangle not yet searched is not particularly interesting for small problems. However, in real world problems of considerable dimension, there might exist hundreds of nondominated basic solutions in a region as the one corresponding to solution 5. Thus, if the objective function values of the involving solutions in the weight space are considered sufficiently close to each other, a great computational effort can be avoided by discarding the search that could be irrelevant from a practical point of view.

The comparison between the two graphs presented in Fig. 4.25 enables to understand the geometry of the nondominated frontier of the problem.

Adjacent nondominated basic solutions (that is, connected by a nondominated edge) have contiguous indifference regions. Thus,

(i) a nondominated vertex in the objective function space corresponds to an indifference region of dimension $p$-1 (polygon) in the weight space;
(ii) a nondominated edge connecting two vertices in the objective function space corresponds to a region of dimension $p$-2 (line segment) in the weight space that belongs to the indifference regions associated with those two vertices;
(iii) a nondominated face in the objective function space corresponds to a point of the weight space (that is, a single strictly positive weight vector) common to all indifference regions associated with the nondominated vertices that define the face.

In this example there are two nondominated faces, one defined by solutions 2, 6, 4 and 5, and another one defined by solutions 1, 3, 5, 4 and 7. The face defined by solutions 4, 6 and 7 is dominated by the edges (4,6) and (4,7), i.e. all the points lying on the face are just weakly nondominated and just the solutions on the edges (4,6) and (4,7) are strictly nondominated. Through the visual inspection of the weight space it is possible to conclude that the weight vector corresponding to this face, (0.429, 0.571, 0), has a zero component in $\lambda_3$. Note that vertices 6 and 7 are adjacent, that is, there is a feasible edge that connects them. Since this edge is weakly nondominated, the variable becoming basic when moving from one extreme of the edge to the other one, although leading to a nondominated vertex, it is not an efficient nonbasic variable (see the notion of efficient nonbasic variable introduced in Chap. 3).

For a better understanding of the geometry of the nondominated region, the three-dimensional objective space with the identification of all nondominated basic solutions is displayed in Fig. 4.26.

In Fig. 4.27 the complete weight space decomposition is presented. Note that the weight space projection presented in Fig. 4.25 provides similar information to the representation in Fig. 4.27, but changing the relative percentages of the areas of the indifference regions.

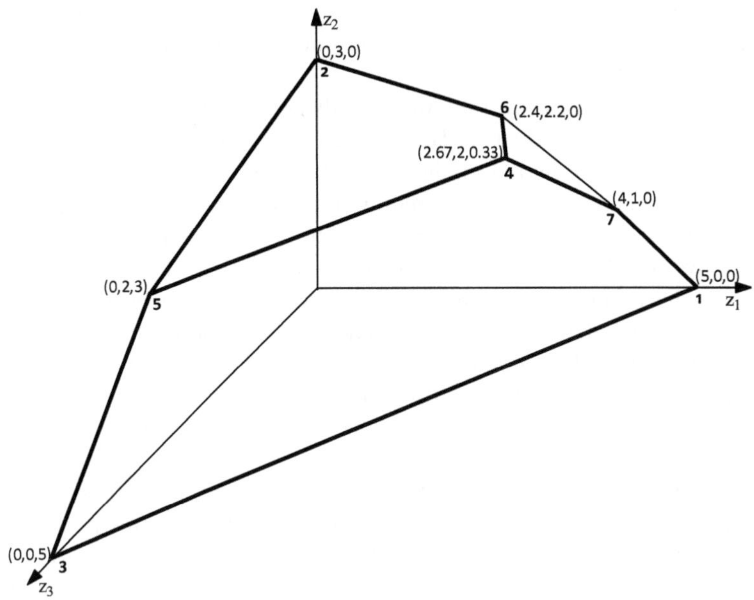


**Fig. 4.26** Identification of nondominated vertices, edges and faces in the objective function space

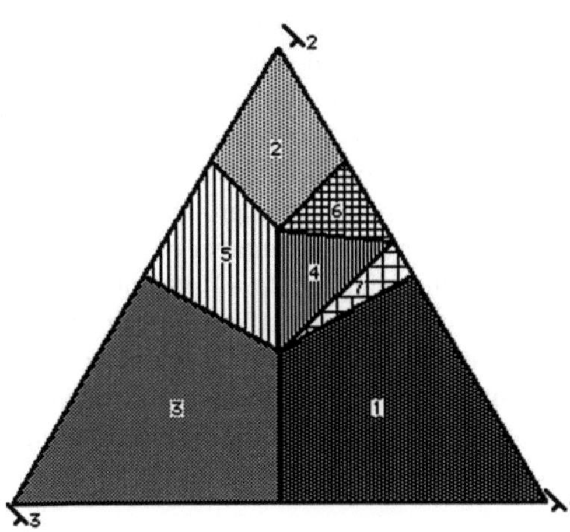


**Fig. 4.27** Representation of the weight space whose projection is displayed in Fig. 4.25

## 4.5 Interval Criterion Weights Method

### *4.5.1 Introduction*

The Interval Criterion Weights (ICW) developed by Steuer (1977, 1986) is an interactive method, which progressively reduces the criterion cone (convex cone generated by the objective function gradients). This reduction is performed according to the DM's preferences, by choosing the solution that he/she prefers from a nondominated solution sample that is presented in each dialogue phase. In each computation phase several weighted sums of the objective functions are optimized.

The ICW method uses several uniformly dispersed vectors in the weight space, to define a set of weighted-sums. Hence, the explicit indication of weight vectors depending on the information elicited from the DM about his/her preferences is avoided. Also, this method does not require the specification of satisfactory values for the objective functions (as, for example, the STEM method). The information provided by the DM by choosing his/her most preferred solution is used for contracting the criterion cone around the objective function used for computing that solution, which determines the reduced criterion cone for the next interaction. The criterion cone is thus gradually reduced around specific directions, until the search is focussed on a small portion of the feasible region, which hopefully contains the nondominated vertex with the highest value for the DM's implicit utility function.

The foundations used for reducing the criterion cone are exposed in detail in Steuer (1986) and are beyond the purpose of this book since this is not a specific issue of multiobjective programming. Therefore, in this text we opted to present the ICW method without detailing the computation of the **T** matrices used in the criterion cone contraction procedure.

The information required from the DM in the dialogue phases concerns his/her preferences among the proposed alternatives obtained in the previous computation phase. The computation of the nondominated solutions proposed to the DM are based on the optimization of weighted-sums of the objective functions, using in each iteration $2p+1$ weight vectors well distributed in the weight space ($p$ being the number of objective functions). The procedure for determining these weight vectors is presented during the description of the algorithm. It should also be noted that the number of nondominated solutions proposed to the DM in each dialogue phase is fixed and may be lower than the number of solutions obtained through the optimization of weighted-sums using the $2p+1$ weight vectors. For this purpose, a filtering procedure may be used to select a sample of $Q$ points, considered as the most different ones, from a maximum of $2p+1$ possible distinct solutions computed in each iteration. The intention is to avoid the comparison by the DM of a high number of solutions. For further details on the diverse filtering techniques that can be used see Steuer (1986, chapter 9).

### 4.5.2 *ICW (Interval Criterion Weights) Algorithm*

A simplified version of the ICW method is presented, regarding the one presented in Steuer (1986).

**Step 1**
Initially, some parameters of the algorithm are specified (following the indications given in Steuer 1986, chapter 13). The DM is asked to specify:

- the sample size, $Q$, of the nondominated solutions being presented in each dialogue phase ($p \leq Q < 2p + 1$);
- the number of iterations $I$ ($I \approx p$).

Steuer also suggests the previous normalization of the objectives (to take into account the different orders of magnitude) by multiplying the coefficients of the decision variables in each objective function by an appropriate power of 10 (instead of using any norm).

Let $h = 1$ (iteration counter).

**Step 2**
In each iteration $2p + 1$ convex combinations of the objective function gradients are formed, with the purpose of obtaining well dispersed nondominated vertices (regarding the criterion cone of the current iteration).

The convex combinations of the objective function gradients are given by the following set of weight vectors:

$\lambda^1 = (1, 0, \ldots, 0)$ Extreme convex combinations (associated with the optimum of each objective function)

$\lambda^2 = (0, 1, \ldots, 0)$

.........

$\lambda^p = (0, 0, \ldots, 1)$

$\lambda^{p+1} = \left(\frac{1}{p^2}, r, \ldots, r\right)$ Non-central convex combinations

$\lambda^{p+2} = \left(r, \frac{1}{p^2}, \ldots, r\right)$

.........

$\lambda^{2p} = \left(r, r, \ldots, \frac{1}{p^2}\right)$

$\lambda^{2p+1} = \left(\frac{1}{p}, \frac{1}{p}, \ldots, \frac{1}{p}\right)$ Central convex combination

and $r = \dfrac{p+1}{p^2}$

In the extreme convex combinations 1-ε is used instead of 1 (for the weights equal to 1), and $0 + \frac{\varepsilon}{p-1}$ instead of zero (for the weights equal to zero). Vectors on the frontier of the weight space are avoided in order to ensure that (strictly) nondominated solutions are computed.

**Step 3**
In each calculation phase the following $2p+1$ linear problems are solved:

$$\max_{\mathbf{x}\in X}\left\{\boldsymbol{\lambda}^k \mathbf{C}_h \mathbf{x}\right\} \quad k = 1, \ldots, 2p+1$$

and

$$\mathbf{C}_h = \begin{cases} \mathbf{T}_{h-1}\ \mathbf{T}_{h-2}\ \ldots\ \mathbf{T}_1\ \mathbf{C} & \text{for} \quad h > 1 \\ \mathbf{C} & \text{for} \quad h = 1 \end{cases}$$

that is, in the initial iteration, $h=1$, the objective function coefficient matrix is the initial matrix **C**, in general normalized (that is, transforming **C** into $\mathbf{C}' = \mathbf{D}\,\mathbf{C}$, where **D** is a diagonal matrix, of $p \times p$ dimension, whose components are the normalization factors of each row of **C**). Note that the normalizing operation does not change the criterion cone, although it influences the weighted objective function gradient. Therefore, it also influences the nondominated solutions resulting from each weighted sum optimization. There is a **T** matrix defined for each weight combination and the pre-multiplication of **C** by the **T** matrices leads to the objective function coefficients in the new iteration.

**Step 4**
The nondominated solutions obtained by optimizing the $2p+1$ weighted sums of the objective functions computed in step 3 are filtered to obtain a sample of size $Q$ and the DM is asked to choose the most satisfactory one according to his/her preferences.

The filtering process has the purpose of selecting the *Q most distinct* points for integrating the solution sample. A specific technique is used. For further details see Steuer (1986, chapter 9), where the following filtering techniques are suggested: *closest point outside the neighborhoods* and *furthest point outside the neighborhoods*.

Then, the objective function coefficient matrix for the next iteration is built by multiplying the previous one by a **T** matrix, leading to a new contracted and dislocated gradient cone with respect to the previous iteration cone. The new contracted cone tends to be centered around the convex combination of the objective function gradients associated with the solution preferred by the DM in the last iteration. This cross-section of the contracted cone is $\frac{1}{p}$ of the cross-section of the previous cone.

If the number $I$ of iterations specified by the DM was not yet performed, then set $h \leftarrow h+1$ and return to step 3.

Otherwise, go to step 5.

**Step 5**
Compute all nondominated vertices not yet known that can be generated using the current cone of the objective functions.

**Step 6**
A sample of the previously computed vertices with dimension $Q$ is determined through a filtering technique. Finally, the DM is asked to choose one of these solutions.

Since the criterion cone is being gradually contracted in each iteration, according to the DM's preferences, the number of reachable nondominated vertices is progressively smaller. Steuer (1986) suggests that after the pre-specified number of $I$ iterations, an algorithm may be applied to compute all nondominated vertices considering the current objective matrix, $\mathbf{C}_I$ (step 5). These solutions are then filtered once more to present a final sample to the DM, allowing him/her to make the choice of the final solution (step 6).

The block diagram of the ICW method is presented in Fig. 4.28.

### 4.5.3 *Final Comments*

As this method uses weighted-sums of the objective functions with weight vectors well dispersed in the weight space, it has the advantage of calculating, in each iteration, potentially well differentiated solutions. However, the method is too rigid in certain issues. For example, it sets *a priori* the maximum number of iterations and does not allow the re-evaluation of decisions made in previous dialogue phases. These issues have the purpose of limiting the computational burden. Also, it may not be easy for the DM (from the cognitive point of view) to select a solution, in each iteration.

Finally, it should also be noted that this method was built with the assumption that there is a DM's implicit utility function, trying to guarantee the convergence to the nondominated vertex that is closer to the optimum of that implicit function.

### 4.5.4 *Illustrative Example of the* Interval Criterion Weights *Method*

Consider the following linear programming problem with three objective functions:

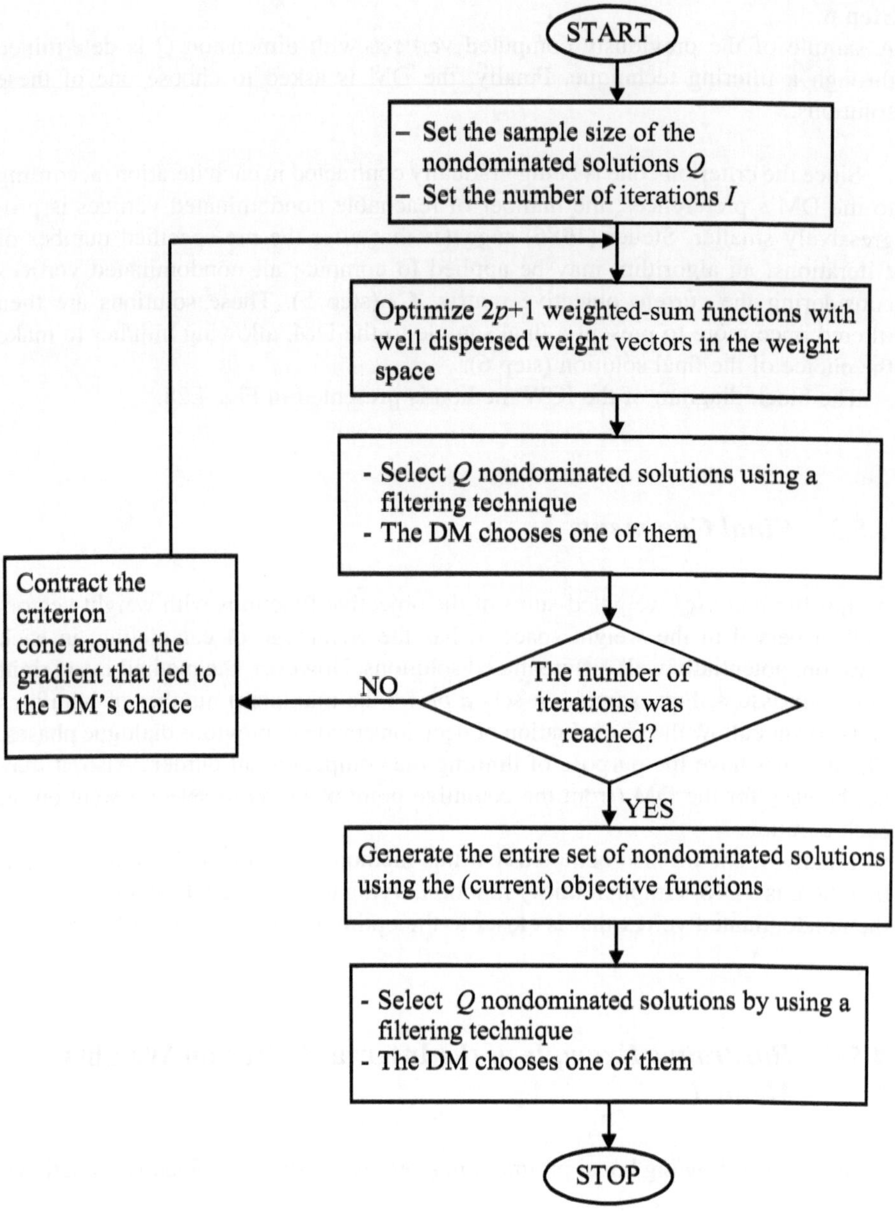


**Fig. 4.28** Block diagram of the ICW method

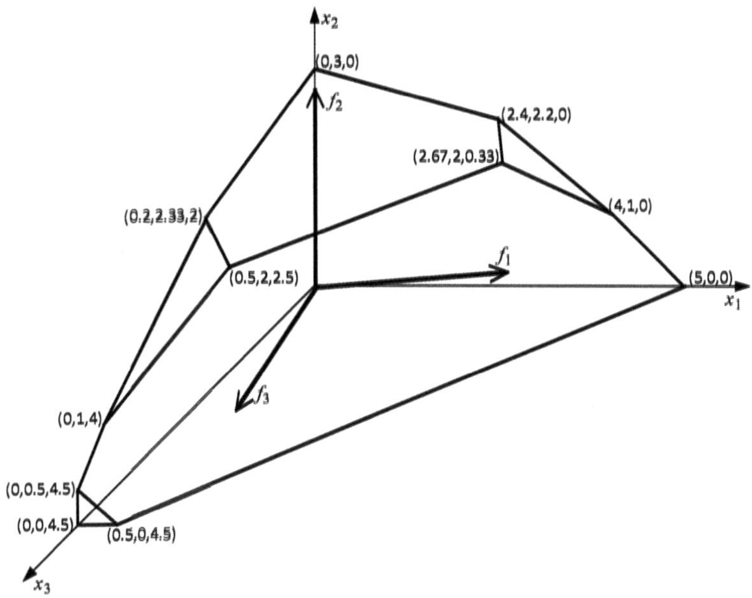


**Fig. 4.29** Feasible region in the decision space

$$
\begin{aligned}
\max \quad & z_1 = f_1(\mathbf{x}) = 5x_1 + x_2 + x_3 \\
\max \quad & z_2 = f_2(\mathbf{x}) = 4.5x_2 \\
\max \quad & z_3 = f_3(\mathbf{x}) = x_1 + 4\ x_3 \\
\text{s.t.} \quad & \left.\begin{aligned}
x_1 + x_2 + x_3 &\le 5 \\
x_1 + 3x_2 + x_3 &\le 9 \\
3x_1 + 4x_2 &\le 16 \\
3x_2 + 2x_3 &\le 11 \\
x_3 &\le 4.5 \\
x_1, x_2, x_3 &\ge 0
\end{aligned}\right\} \text{feasible region } X
\end{aligned}
$$

Figure 4.29 shows the feasible region $X$ in the decision space, where the gradients of the 3 objective functions ($f_1$,$f_2$,$f_3$) are also displayed (in a different scale w.r.t. $X$).

Step 1. Initially, the DM is asked to specify the sample size concerning the number of nondominated solutions to be presented, $Q$, and the number of iterations, $I$. Steuer (1986) suggests that $Q$ should be between $p$ (the number of objective functions) and $2p+1$ (number of convex combinations of the objective function gradients used in each iteration for computing nondominated solutions).

Suppose that the DM wants to know all the solutions computed in each iteration. In this way, $Q = 2p + 1 = 7$, and no filtering process is necessary.

Consider the number of iterations $I = 3$.

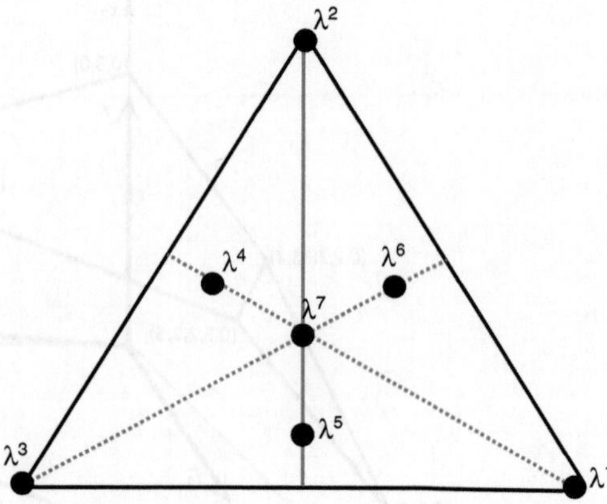


**Fig. 4.30** Weight vectors well dispersed in the weight space

Step 2. A number of $2p+1=7$ well dispersed weight vectors is built (Fig. 4.30). The corresponding convex combinations of the gradients of the objective functions are:

$$\boldsymbol{\lambda}^1 = (1,\ 0,\ 0) \qquad \text{(Extreme convex combinations)}$$
$$\boldsymbol{\lambda}^2 = (0,\ 1,\ 0)$$
$$\boldsymbol{\lambda}^3 = (0,\ 0,\ 1)$$
$$\boldsymbol{\lambda}^4 = \left(\frac{1}{p^2},\ r,\ r\right) = (0.111,\ 0.444,\ 0.444) \qquad \text{(Non-central convex combinations)}$$
$$\boldsymbol{\lambda}^5 = \left(r,\ \frac{1}{p^2},\ r\right) = (0.444,\ 0.111,\ 0.444)$$
$$\boldsymbol{\lambda}^6 = \left(r,\ r,\ \frac{1}{p^2}\right) = (0.444,\ 0.444,\ 0.111)$$
$$\boldsymbol{\lambda}^7 = \left(\frac{1}{p},\ \frac{1}{p},\ \frac{1}{p}\right) = (0.333,\ 0.333,\ 0.333) \qquad \text{(Central convex combination)}$$

where $r = \dfrac{p+1}{p^2} = \dfrac{4}{9} = 0.444.$

**First Iteration**

Step 3. In the first iteration, the objective function matrix $\mathbf{C}_1$ is the initial matrix $\mathbf{C}$:

$$\mathbf{C}_1 = \mathbf{C} = \begin{bmatrix} 5 & 1 & 1 \\ 0 & 4.5 & 0 \\ 1 & 0 & 4 \end{bmatrix}$$

It is not necessary to normalize the objective function coefficients in this problem because they already have the same order of magnitude.

The following seven weighted-sum problems are solved, $\left(P_{\lambda^k}\right)$:

$$\max_{\mathbf{x}\in X} \boldsymbol{\lambda}^k \mathbf{C}_1 \mathbf{x} \quad k = 1, \ldots, 7$$

The first instance to solve $(P_{\lambda^1})$ is:

$$\begin{aligned} &\max \quad \begin{bmatrix} 1-\varepsilon & \frac{\varepsilon}{2} & \frac{\varepsilon}{2} \end{bmatrix} \begin{bmatrix} 5 & 1 & 1 \\ 0 & 4.5 & 0 \\ 1 & 0 & 4 \end{bmatrix} \begin{bmatrix} x_1 \\ x_2 \\ x_3 \end{bmatrix} \\ &\text{s.t.} \quad \mathbf{x} \in X \end{aligned}$$

with ε a small positive value, e.g. 0.001.

The optimal solution to this problem (which is a nondominated solution to the multiobjective problem) is solution 1: $\mathbf{x}^1 = (5, \ 0, \ 0)$, $\mathbf{z}^1 = (25, \ 0, \ 5)$.

Using the remaining weight vectors $\boldsymbol{\lambda}^2, \ldots, \boldsymbol{\lambda}^7$ the remaining six instances $(P_{\lambda^k})$ are solved. The following nondominated solutions are obtained:

$$\begin{array}{lll} (P_{\lambda^2})\text{-solution 2} & \mathbf{x}^2 = (0, \ 3, \ 0) & \mathbf{z}^2 = (3, \ 13.5, \ 0) \\ (P_{\lambda^3})\text{-solution 3} & \mathbf{x}^3 = (0.5, \ 0, \ 4.5) & \mathbf{z}^3 = (7, \ 0, \ 18.5) \\ (P_{\lambda^4})\text{-solution 4} & \mathbf{x}^4 = (0, \ 1, \ 4) & \mathbf{z}^4 = (5, \ 4.5, \ 16) \\ (P_{\lambda^5})\text{-solution 1} & \mathbf{x}^1 = (5, \ 0, \ 0) & \mathbf{z}^1 = (25, \ 0, \ 5) \\ (P_{\lambda^6})\text{-solution 5} & \mathbf{x}^5 = (4, \ 1, \ 0) & \mathbf{z}^5 = (21, \ 4.5, \ 4) \\ (P_{\lambda^7})\text{-solution 1} & \mathbf{x}^1 = (5, \ 0, \ 0) & \mathbf{z}^1 = (25, \ 0, \ 5) \end{array}$$

Although seven linear problems have been solved, only five distinct nondominated solutions were obtained. Note that solving $(P_{\lambda^5})$, $(P_{\lambda^7})$ and $(P_{\lambda^1})$ led to the same solution.

Figure 4.31 presents the indifference regions of these solutions on the weight space.

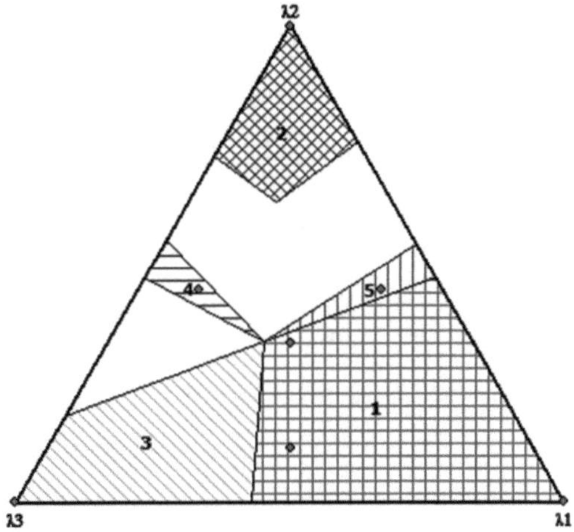


**Fig. 4.31** Indifference regions of the solutions computed in the first iteration

Step 4. The solutions computed in step 3 are presented to the DM, allowing him/her to choose the *best one* according to his/her preferences.

Suppose that the DM chooses solution 4, obtained using the convex combination of the objective function gradients associated with the weight vector $\boldsymbol{\lambda}^4 = (\ 0.111,\ \ 0.444,\ \ 0.444\ )$.

The matrix corresponding to this weight vector is $\mathbf{T}^{(4)}$, which is used to obtain the criterion cone of the next iteration, contracted and dislocated from the cone of the previous iteration.

The objective matrix of the second iteration is obtained from the matrix of the current iteration by making $\mathbf{C}_2 = \mathbf{T}_1\mathbf{C}_1$, where $\mathbf{T}_1 = \mathbf{T}^{(4)}$.

Matrix $\mathbf{T}_1$ is obtained such that the new criterion cone is contracted around $\boldsymbol{\lambda}^4\mathbf{C}_1$. As it was already mentioned, there is a pre-defined $\mathbf{T}$ matrix for each of the $2p+1$ convex combinations (for technical details see (Steuer 1986, chapter 9)):

$$\boldsymbol{\lambda}^1 : \mathbf{T}^{(1)} = \begin{bmatrix} 1 & 0 & 0 \\ q & 1-q & 0 \\ q & 0 & 1-q \end{bmatrix}$$

$$\boldsymbol{\lambda}^2 : \mathbf{T}^{(2)} = \begin{bmatrix} 1-q & q & 0 \\ 0 & 1 & 0 \\ 0 & q & 1-q \end{bmatrix} \qquad \text{(extreme convex combinations)}$$

$$\boldsymbol{\lambda}^3 : \mathbf{T}^{(3)} = \begin{bmatrix} 1-q & 0 & q \\ 0 & 1-q & q \\ 0 & 0 & 1 \end{bmatrix}$$

$$\boldsymbol{\lambda}^4 : \mathbf{T}^{(4)} = \begin{bmatrix} 1-q & \frac{q}{a} & \frac{q}{a} \\ 0 & 1-(p-2)\frac{q}{a} & \frac{q}{a} \\ 0 & \frac{q}{a} & 1-(p-2)\frac{q}{a} \end{bmatrix}$$

$$\boldsymbol{\lambda}^5 : \mathbf{T}^{(5)} = \begin{bmatrix} 1-(p-2)\frac{q}{a} & 0 & \frac{q}{a} \\ \frac{q}{a} & 1-q & \frac{q}{a} \\ \frac{q}{a} & 0 & 1-(p-2)\frac{q}{a} \end{bmatrix} \qquad \text{(non-central convex combinations)}$$

$$\lambda^6 : \mathbf{T}^{(6)} = \begin{bmatrix} 1-(p-2)\frac{q}{a} & \frac{q}{a} & 0 \\ \frac{q}{a} & 1-(p-2)\frac{q}{a} & 0 \\ \frac{q}{a} & \frac{q}{a} & 1-q \end{bmatrix}$$

$$\lambda^7 : \mathbf{T}^{(7)} = \begin{bmatrix} 1-q\frac{a}{p} & \frac{q}{p} & \frac{q}{p} \\ \frac{q}{p} & 1-q\frac{a}{p} & \frac{q}{p} \\ \frac{q}{p} & \frac{q}{p} & 1-q\frac{a}{p} \end{bmatrix} \quad \text{(central convex combination)}$$

where $a = p - 1$ and $q = 1 - p^{-\frac{1}{a}}$

For $p = 3$, it is obtained:

$$\mathbf{T}_1 \equiv \mathbf{T}^{(4)} = \begin{bmatrix} 0.577 & 0.211 & 0.211 \\ 0 & 0.789 & 0.211 \\ 0 & 0.211 & 0.789 \end{bmatrix}$$

**Second Iteration**

Step 3. The objective function matrix corresponding to the contracted cone is given by[3]:

$$\mathbf{C}_2 = \mathbf{T}_1\ \mathbf{C}_1 = \begin{bmatrix} 0.577 & 0.211 & 0.211 \\ 0 & 0.789 & 0.211 \\ 0 & 0.211 & 0.789 \end{bmatrix} \begin{bmatrix} 5 & 1 & 1 \\ 0 & 4.5 & 0 \\ 1 & 0 & 4 \end{bmatrix}$$

$$= \begin{bmatrix} 3.098 & 1.528 & 1.423 \\ 0.211 & 3.549 & 0.845 \\ 0.789 & 0.951 & 3.155 \end{bmatrix}$$

In Fig. 4.32 the solutions already known are labelled and the objective function gradients of the contracted cone regarding matrix $\mathbf{C}_2$ (corresponding to vectors $f_1^2, f_2^2$ and $f_3^2$) are also represented.

The following seven weighted-sum problems are solved, $(P_{\lambda^k})$:

[3] In this example the result of the multiplication of matrices is slightly different from the right hand side of the equality. These discrepancies are due to the fact that the computations were performed with a higher precision.

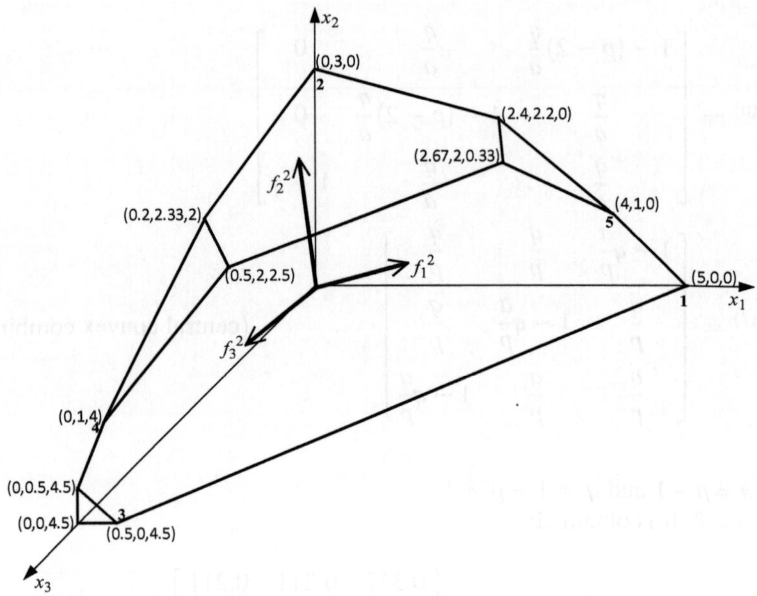


**Fig. 4.32** Objective function gradients of the contracted cone in the second iteration

$$\max_{\mathbf{x}\in X} \; \boldsymbol{\lambda}^k \mathbf{C}_2 \mathbf{x} \quad k = 1, \ldots, 7$$

The solutions obtained are:

| | | |
|---|---|---|
| $(P_{\lambda^1})$-solution 1 | $\mathbf{x}^1 = (5,\ 0,\ 0)$ | $\mathbf{z}^1 = (25,\ 0,\ 5)$ |
| $(P_{\lambda^2})$-solution 2 | $\mathbf{x}^2 = (0,\ 3,\ 0)$ | $\mathbf{z}^2 = (3,\ 13.5,\ 0)$ |
| $(P_{\lambda^3})$-solution 6 | $\mathbf{x}^6 = (0,\ 0.5,\ 4.5)$ | $\mathbf{z}^6 = (5,\ 2.24,\ 18)$ |
| $(P_{\lambda^4})$-solution 4 | $\mathbf{x}^4 = (0,\ 1,\ 4)$ | $\mathbf{z}^4 = (5,\ 4.5,\ 16)$ |
| $(P_{\lambda^5})$-solution 3 | $\mathbf{x}^3 = (0.5,\ 0,\ 4.5)$ | $\mathbf{z}^3 = (7,\ 0,\ 18.5)$ |
| $(P_{\lambda^6})$-solution 7 | $\mathbf{x}^7 = (2.667,\ 2,\ 0.333)$ | $\mathbf{z}^7 = (15.667,\ 9,\ 4)$ |
| $(P_{\lambda^7})$-solution 4 | $\mathbf{x}^4 = (0,\ 1,\ 4)$ | $\mathbf{z}^4 = (5,\ 4.5,\ 16)$ |

Figure 4.33 shows the indifference regions associated with these solutions:

(a) regarding the weight space corresponding to the contracted cone;
(b) regarding the original weight space, where the weight space of the contracted cone can also be seen (shaded background) overlapping the original one.

Step 4. The solutions computed in step 3 are presented to the DM, allowing him/her to choose the most satisfactory regarding his/her preferences.

Suppose that the DM chooses solution 6, which is obtained using the convex combination of the objective function gradients of the contracted cone with the

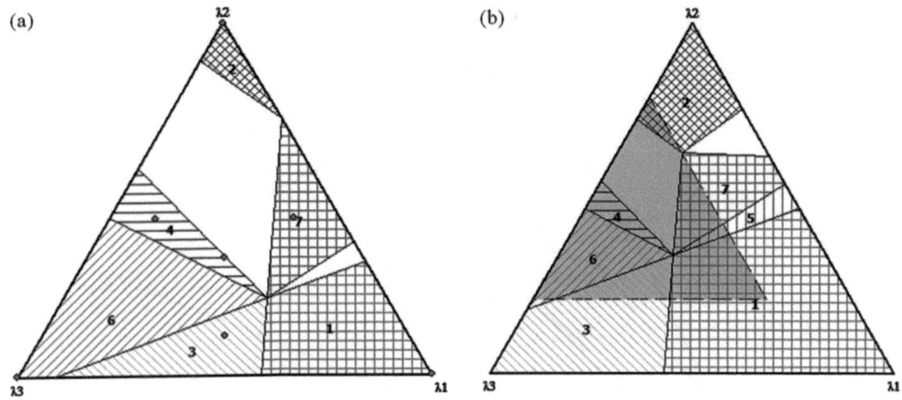


**Fig. 4.33** Indifference regions of the nondominated solutions obtained in the second iteration. (**a**) Weight space corresponding to the contracted cone. (**b**) Original weight space (the weight space corresponding to the contracted cone is displayed with shaded background)

weight vector $\boldsymbol{\lambda}^3 = (\varepsilon/2,\ \varepsilon/2,\ 1-\varepsilon)$. With this information it is possible to obtain the contraction matrix of the second iteration, $\mathbf{T}_2$:

$$\mathbf{T}_2 = \mathbf{T}^{(3)} = \begin{bmatrix} 0.577 & 0 & 0.423 \\ 0 & 0.577 & 0.423 \\ 0 & 0 & 1 \end{bmatrix}$$

The objective function coefficients matrix for the next iteration is $\mathbf{C}_3 = \mathbf{T}_2\ \mathbf{C}_2 = \mathbf{T}_2\ \mathbf{T}_1\ \mathbf{C}$, leading to a new contracted and dislocated cone regarding the previous one.

**Third Iteration**

Step 3. The objective function matrix $\mathbf{C}_3$, corresponding to the contracted cone is given by:

$$\mathbf{C}_3 = \mathbf{T}_2\ \mathbf{C}_2 = \begin{bmatrix} 0.577 & 0 & 0.423 \\ 0 & 0.577 & 0.423 \\ 0 & 0 & 1 \end{bmatrix} \begin{bmatrix} 3.098 & 1.528 & 1.423 \\ 0.211 & 3.549 & 0.845 \\ 0.789 & 0.951 & 3.155 \end{bmatrix}$$

$$= \begin{bmatrix} 2.122 & 1.284 & 2.155 \\ 0.455 & 2.451 & 1.821 \\ 0.789 & 0.951 & 3.155 \end{bmatrix}$$

In Fig. 4.34 the solutions already known are labelled, and the gradients of the contracted cone of the objective functions are also displayed, regarding matrix $\mathbf{C}_3$ (corresponding to vectors $f_1^3, f_2^3$ and $f_3^3$).

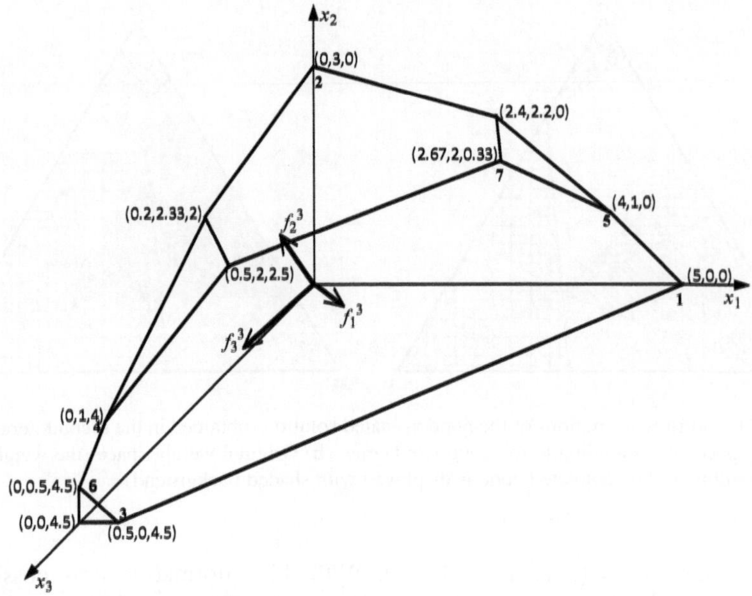


**Fig. 4.34** Objective function gradients of the contracted cone in the third iteration

The following seven weighted-sum problems are solved, $(P_{\lambda^k})$:

$$\max_{\mathbf{x}\in X} \boldsymbol{\lambda}^k \mathbf{C}_3\mathbf{x} \quad k = 1, \ldots, 7$$

The nondominated solutions found are:

$(P_{\lambda^1})$-solution 3
$(P_{\lambda^2})$-solution 4
$(P_{\lambda^3})$-solution 6
$(P_{\lambda^4})$-solution 6
$(P_{\lambda^5})$-solution 3
$(P_{\lambda^6})$-solution 6
$(P_{\lambda^7})$-solution 6

All these solutions were already known in the previous iterations. However, it should be noted that there are nondominated vertices of the original problem not yet computed until this phase of the search (Fig. 4.35b).

Figure 4.35 presents the indifference regions of these solutions both regarding the weight space corresponding to the contracted cone and the original weight space (where the weight space of the contracted cone is shown with a shaded background overlapping the original one).

Since three iterations were already performed, the iterative process stops.

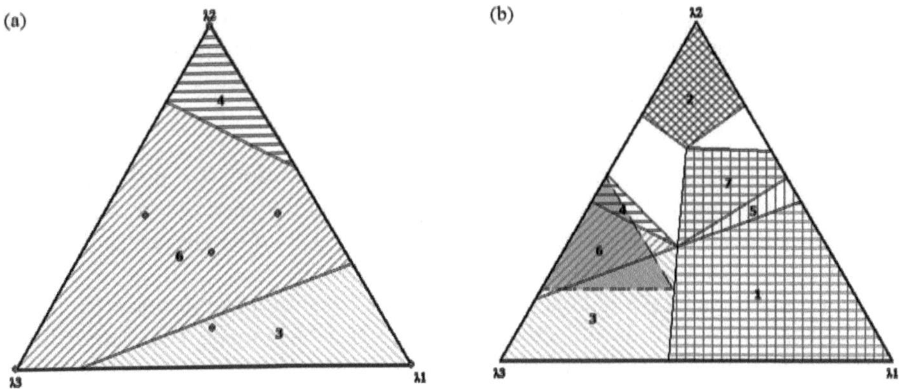

**Fig. 4.35** Indifference regions of the solution in the third iteration. (**a**) Weight space corresponding to the contracted cone. (**b**) Original weight space

Steuer (1986) suggests that, in the end, an algorithm should be applied to compute all the nondominated basic solutions considering the objective functions of the final contracted criterion cone (matrix $\mathbf{C}_3$ —Step *5*). Nevertheless, in this case, no additional computations are required. As it can be seen in Fig. 4.35, the indifference regions of solutions 3, 4 and 6 completely fill the weight space corresponding to the contracted cone, meaning that these solutions form the entire set of nondominated basic solutions that can be obtained with matrix $\mathbf{C}_3$.

Step 6. In this case, the number of solutions obtained does not justify the application of a filtering process. Then, the DM should choose a compromise solution among solutions 3, 4 and 6.

## 4.6 Pareto Race Method

### *4.6.1 Method Description*

The Pareto Race method proposed by Korhonen and Wallenius (1988) is based on the work developed by Korhonen (1987) and Korhonen and Laakso (1986a, 1986b). It is a directional search method allowing the DM moving freely on the nondominated region. The information required from the DM essentially consists in the specification of the objective functions to be improved, which changes the direction of the motion. The nondominated solutions are obtained through the optimization of a reference point scalarizing function and the use of parametric programming.

A reference direction is built from the aspiration levels for the objective function values, initially specified by the DM. This direction starts from a point in the objective function space and offers a variation of the values of the objective functions according to the DM's preferences. The reference direction is then projected onto the nondominated solution set by using an achievement scalarizing function, which corresponds to a weighted Chebyshev metric if the reference direction is not attainable, generating a trajectory (sub-set of the nondominated solutions) that is presented to the DM. In this way, the DM can travel on the nondominated frontier, controlling the direction of motion (by privileging the objective functions that he/she wants to improve) and the speed (controlling how close on that direction solutions should be calculated), as if he/she is driving a car (hence, the designation Pareto Race) on that surface.

According to generalized goal programming, a constraint can be considered as a non-flexible target and an objective function a flexible target. The Pareto Race method considers the set $G$ of flexible targets, which are associated with the aspiration levels of the objective functions, and the set $R$ of the non-flexible targets (constraints).

Initially the DM is asked to specify the values of the aspiration levels for the objective functions, i.e. the starting reference point. From this data, the vector $\mathbf{b}$ is obtained, with dimension $m+p$, including the values of the right-hand sides of the constraints ($m$ non-flexible targets) and the aspiration levels for each objective function ($p$ flexible targets).

The range of variation of the flexible targets ($\Delta b_k$) is also specified by the DM. These values implicitly limit the relative importance of each objective function:

$$w_k = \begin{cases} \Delta b_k & \text{if} \quad k \in G \\ 0 & \text{if} \quad k \in R \end{cases}$$

$\mathbf{d}$ is the reference vector which controls the direction of motion. Initially $\mathbf{d} = \mathbf{w}$.

The normalization of the directions is achieved by maintaining $s$ constant in all iterations: $s = \sum_k d_k$.

In order to compute nondominated solutions resulting from the projection of the unbounded line segment, $\mathbf{b} + t\,\mathbf{d}$, the following linear parametric problem is solved:

$$\begin{array}{lll} \min & \left\{ v - \rho \sum_{k=1}^{p} \mathbf{c}_k \mathbf{x} \right\} & \\ \text{s.t.} & \mathbf{c}_j\ \mathbf{x} + v\ w_j \geq b_j + t\ d_j & j \in G \\ & \mathbf{A}_{i\bullet}\ \mathbf{x} = b_i & i \in R \\ & \mathbf{x} \geq \mathbf{0} & \\ & v \in \mathbb{R} & \end{array} \tag{4.4}$$

where ρ is a very small positive number and $(\rho \sum_{k=1}^{p} \mathbf{c}_k \mathbf{x})$ is a perturbation term aimed at enforcing the computation of a nondominated solution (when there are alternative optima for (4.4)). By omitting this term, there is only the guarantee that the solution obtained is weakly nondominated. The vectors $\mathbf{A}_{i\bullet}$, $i = 1,\ldots,m$, contain the decision variable coefficients in the constraints (non-flexible targets), as well as the coefficients associated with the auxiliary variables used for converting the inequalities into equalities. The constraints $\mathbf{A}_{i\bullet}\ \mathbf{x} = b_i,\ i \in R$, are equivalent to $\mathbf{A}_{i\bullet}\ \mathbf{x} = b_i + t\ d_i$, with $d_i = 0$ for $i \in R$. Note that the variable $v$ can be positive or negative, since the reference point can either be inside or outside the feasible region. In order to deal with non-negative variables only, $v$ is replaced by $v = v^+ - v^-$, with $v^+, v^- \geq 0$.

The parameter $t$ controls the speed of motion.

Therefore, the projection of the reference point onto the nondominated region is obtained by minimizing an achievement scalarizing function. The computation of the trajectory on the nondominated region is made by using a parametric programming problem regarding the right-hand sides of the constraints associated with the objective functions of the MOLP problem. Let the reference point be $\mathbf{b}_G$ (i.e. the part of vector $\mathbf{b}$, such that its components are $b_j, j \in G$) and $\boldsymbol{\lambda}$ the weight vector (note that, in this setting, the weights are essentially scaling factors). The scalarizing program is:

$$\begin{aligned} \min\ & \left\{ \max_{j=1,\cdots,p} \left\{ \lambda_j \left( b_j - \mathbf{c}_j\mathbf{x} \right) \right\} - \rho \sum_{k=1}^{p} \mathbf{c}_k\mathbf{x} \right\} \\ \text{s.t.}\ & \mathbf{A}_{i\bullet}\mathbf{x} = b_i \quad i \in R \\ & \mathbf{x} \geq \mathbf{0} \end{aligned}$$

which is equivalent to:

$$\begin{aligned} \min\ & \left\{ v - \rho \sum_{k=1}^{p} \mathbf{c}_k\mathbf{x} \right\} \\ \text{s.t.}\ & \mathbf{c}_j\mathbf{x} + \frac{v}{\lambda_j} \geq b_j \quad j \in G \\ & \mathbf{A}_{i\bullet}\mathbf{x} = b_i \qquad i \in R \\ & \mathbf{x} \geq \mathbf{0}, \quad v \in \mathbb{R} \end{aligned}$$

This problem can be written as:

$$\begin{aligned} \min\ & \left\{ v - \rho \sum_{k=1}^{p} \mathbf{c}_k\mathbf{x} \right\} \\ \text{s.t.}\ & \lambda_j \left( b_j - \mathbf{c}_j\mathbf{x} \right) \leq v \quad j \in G \\ & \mathbf{A}_{i\bullet}\mathbf{x} = b_i \qquad i \in R \\ & \mathbf{x} \geq \mathbf{0} \\ & v \in \mathbb{R} \end{aligned} \tag{4.5}$$

This problem is similar to the one solved in the STEM method, except the term $\rho \sum_{k=1}^{p} \mathbf{c}_k\mathbf{x}$ in the objective function and $v \in \mathbb{R}$ because the reference point may be reachable. That term can also be introduced in the STEM method with the same purpose, i.e. enforcing obtaining a strictly nondominated solution.

It can be seen that problem (4.5) is different from (4.4) because the right-hand sides of the constraints in (4.5) are not parameterized. In order to define the direction towards which the reference point should be moving, $td_j$ is added to $b_j$ for $j \in G$ ($d_j$ is the component $j$ of the direction and $t$ the parameter that controls the movement). The weight $\lambda_j$ has a similar meaning to the $\alpha_j$ used in the STEM method. In the Pareto Race method the weight $w_j$ is used, being defined by $w_j = \frac{1}{\lambda_j}$.

Each time that (4.4) is solved, $t = 0$ is initially considered. Using classical sensitivity analysis regarding the right-hand sides of the constraints of (4.4), for the optimal solution associated with a certain value of $t$, in particular $t = 0$, the interval $[t - t_1, \; t + t_2]$ is computed, indicating the possible variation of $t$ in order that the current basis remains feasible.

Solving the problem (4.4) for a particular value of $t$ using the simplex method, the values of the basic variables in the optimal solution are given by:

$$\mathbf{x}_B = \mathbf{B}^{-1}(\mathbf{b} \; + \; t \; \mathbf{d})$$

where $\mathbf{x}_B$ is the vector of basic variables and $\mathbf{B}^{-1}$ the inverse of the basis matrix.

By changing $t$ into $t + \theta$, it is obtained:

$$\mathbf{x}_B(\theta) = \mathbf{B}^{-1}(\mathbf{b} + (t + \theta) \; \mathbf{d})$$
$$\mathbf{x}_B(\theta) = \mathbf{x}_B + \theta \; \mathbf{B}^{-1}\mathbf{d}$$

The upper and lower bounds of $\theta$ are computed such that the solution remains feasible (since the optimality condition is not affected), that is, $\mathbf{x}_B(\theta) \geq \mathbf{0}$. The interval $\theta \in [t - t_1, \; t + t_2]$ is obtained.

From the initial nondominated solution, the DM has several options, which do not request the specification of concrete values but only the indication of the variation trends, i.e.:

- **Proceeding** along the current direction at constant speed. The step is updated by making

$$t \leftarrow \begin{cases} t + min \; \{\Delta t \; ; \; t_2\} & \text{if} \quad \Delta t > 0 \\ t + max \; \{\Delta t \; ; \; -t_1\} & \text{if} \quad \Delta t < 0 \end{cases}$$

  $\Delta t$ represents the speed of motion and it is initially equal to β, i.e. a scalar representing the basic pace.

  If $t_2 = +\infty$, the DM is asked to change the direction.

  If $t_2 = 0$ and $\Delta t > 0$, it means that it is not possible to proceed along the current direction without changing the basis. The necessary operations are performed in order to obtain a new basis and to update the interval $[t - t_1, \quad t + t_2]$.

  If $t_1 = 0$ and $\Delta t < 0$, the DM is asked to change the direction.

If none of these cases occur, the DM should proceed along the same direction with a pace $\Delta t$, the new solution being $\mathbf{x}_B \leftarrow \mathbf{x}_B + \Delta t\mathbf{B}^{-1}\mathbf{d}$. The values $t_1$ and $t_2$ are updated.

- **Changing direction**, in order to improve a certain objective function. The component of the reference direction corresponding to this target is increased:

$$d_j \leftarrow d_j + \sigma_j \Delta b_j \quad j \in G$$

where $j$ is the index of the objective function chosen by the DM and $\sigma_j$ is a scalar allowing to determine the variation in the direction of motion.

The direction vector is re-normalized such that $\sum_k d_k = s$.

The problem (4.4) is solved for $t$ =0, and the interval $[t - t_1, \ t + t_2]$ is recomputed. The $b_j$, $j \in G$, considered in (4.4), when the DM wants to change the direction, are the values of the flexible targets $j$ corresponding to the point of the nondominated frontier where the change of direction occurs.

- **Changing course**, making an opposite motion (forward or backward) on the current direction.

$$\Delta t \leftarrow \begin{cases} \beta & \text{if} \quad \Delta t < 0: \quad \text{inversion of the course forward} \\ -\beta & \text{if} \quad \Delta t > 0: \quad \text{inversion of the course backward} \end{cases}$$

The change of course always starts with speed β, although it is possible to maintain the current speed, i.e. the speed when the motion is reversed ($\Delta t$).

- **Increase or decrease the speed**, moving faster or slower, in the current course and direction.

The increase or decrease in the speed is obtained by varying the absolute value of the scalar $\Delta t$. In order to increase the speed $\Delta t \leftarrow \alpha\ \Delta t$, with $\alpha$ >1, is performed. In order to decrease the speed $\Delta t \leftarrow \frac{\Delta t}{\alpha}$, with $\alpha$ >1, is performed (until reaching a minimum equal to β).

- **"Fixing" the level of an objective function**, by introducing a lower bound into a flexible target equal to the corresponding current value.

The constraint of the type $\mathbf{c}_j\mathbf{x} \geq L_j$ is introduced, where $L_j$ is the lower bound corresponding to the objective function $f_j(\mathbf{x})$, $j \in G$.

- **Release an objective function**, by removing the limitation previously imposed.

The block diagram of the Pareto Race method is shown in Fig. 4.36.

### 4.6.2 *Final Comments*

The use of the Pareto Race method is particularly interesting for the DM from the cognitive point of view. It offers a free search in which the DM decides where to travel on the nondominated region. When there is no previous overview of the

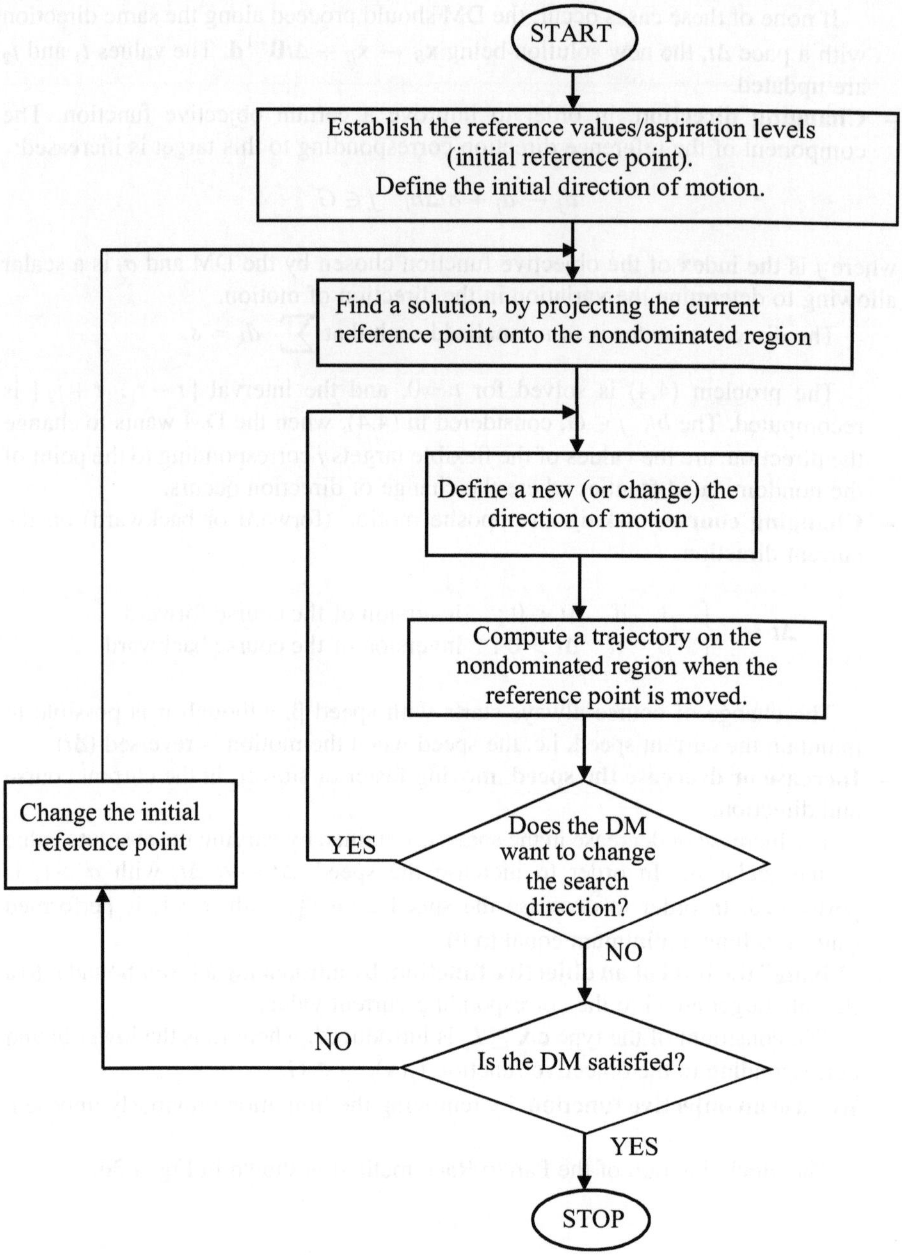


**Fig. 4.36** Block diagram of the Pareto Race method

shape of the feasible region, this method is mostly suited if the trajectories on the nondominated region, chosen by the DM by trial and error, can be performed fast enough. Therefore, the search process of Pareto Race seems particularly suited for a final computation phase, previously considering a strategic search using other procedure(s). The Pareto Race method could then be used to study in more detail the solutions of a delimited nondominated region.

### 4.6.3 Illustrative Example of the Pareto Race Method

Consider the following linear problem with three objective functions (which was used above for illustrating the TRIMAP method):

$$\begin{aligned} \max\ z_1 &= f_1(\mathbf{x}) = x_1 \\ \max\ z_2 &= f_2(\mathbf{x}) = x_2 \\ \max\ z_3 &= f_3(\mathbf{x}) = x_3 \\ \text{s.t.}\ x_1 + x_2 + x_3 &\le 5 \\ x_1 + 3x_2 + x_3 &\le 9 \\ 3x_1 + 4x_2 &\le 16 \\ x_1, x_2, x_3 &\ge 0 \end{aligned}$$

The Pareto Race method starts by asking the DM to specify the aspiration levels and the desirable ranges of variation of the objective function values.

Suppose that the DM establishes the aspiration levels with values 6, 5 and 5 for $f_1(\mathbf{x})$, $f_2(\mathbf{x})$ and $f_3(\mathbf{x})$, respectively, and he/she would like to keep the objective function values in the following intervals: [4.5, 7] for $f_1(\mathbf{x})$, [2.5, 6] for $f_2(\mathbf{x})$ and [2, 6] for $f_3(\mathbf{x})$. Since it may not be possible to obtain the objective function values within these intervals, they are only indicative. In this way, (6, 5, 5) is the original reference point that will be projected onto the nondominated frontier and the proposed variations, (7–4.5, 6–2.5, 6–2) = (2.5, 3.5, 4), form the initial weight vector **w**. Note that the weights have here the role of scale factors. The variation intervals $[LB_j, UB_j]$ for each objective function ($j \in G$) will be updated during the procedure.

The right-hand side vector (*targets*) is $\mathbf{b} = \begin{bmatrix} \mathbf{b}_G \\ -- \\ \mathbf{b}_R \end{bmatrix} = \begin{bmatrix} 6 \\ 5 \\ 5 \\ -- \\ 5 \\ 9 \\ 16 \end{bmatrix}$ and the weight vector is $\mathbf{w} = \begin{bmatrix} 2.5 \\ 3.5 \\ 4 \\ -- \\ 0 \\ 0 \\ 0 \end{bmatrix}$, where the first three components refer to the objectives (flexible targets) and the last three to the constraints (rigid targets). Initially, the reference direction is considered as $\mathbf{d} = \mathbf{w}$.

The directions are normalized by computing the value $s$ for the initial reference direction such that $s = \sum_{j \in G} d_j$, and $s$ is kept constant in all iterations. In this example, $s = 10$.

In order to project the unbounded segment line $\mathbf{b} + t\mathbf{d}$ onto the nondominated solution set, leading to a nondominated trajectory, the parametric linear problem (4.4) is solved.

Problem (4.4) is firstly solved for $t = 0$. The computation of other nondominated solutions for different values of $t$ can be made by using parametric programming regarding the right-hand side of the constraints $G$.

The variable $v$ has no sign restriction and it can be rewritten as $v = v^+ - v^-$ with $v^+, v^- \geq 0$. Thus, the problem to be solved is (considering $\rho = 0.001$):

$$\begin{aligned} \min \quad & v^+ - v^- - 0.001(x_1 + x_2 + x_3) \\ \text{s.t.} \quad & 2.5\, v^+ - 2.5\, v^- + x_1 \geq 6 \\ & 3.5v^+ - 3.5v^- + x_2 \geq 5 \\ & 4v^+ - 4v^- + x_3 \geq 5 \\ & x_1 + x_2 + x_3 \leq 5 \\ & x_1 + 3x_2 + x_3 \leq 9 \\ & 3x_1 + 4x_2 \leq 16 \\ & x_1, x_2, x_3, v^+, v^- \geq 0 \end{aligned}$$

The optimal simplex tableau converted into maximization and omitting the columns of the basic variables is:

| | $\mathbf{c}$ | 1 | 0 | 0 | 0 | 0 | |
|---|---|---|---|---|---|---|---|
| $(\mathbf{c}_B)^T$ | $\mathbf{x}_B$ | $v^-$ | $s_1$ | $s_2$ | $s_3$ | $s_4$ | |
| 0.001 | $x_3$ | 0 | 0.4 | 0.4 | -0.6 | 0.4 | 0.6 |
| 0 | $s_5$ | 0 | -0.7 | 1.3 | -0.7 | -1.7 | 1.7 |
| 0 | $s_6$ | 0 | 0.85 | 1.85 | -2.15 | -2.15 | 1.65 |
| 0.001 | $x_1$ | 0 | -0.75 | 0.25 | 0.25 | 0.25 | 3.25 |
| 0.001 | $x_2$ | 0 | 0.35 | -0.65 | 0.35 | 0.35 | 1.15 |
| -1 | $v^+$ | -1 | -0.1 | -0.1 | -0.1 | -0.1 | 1.1 |
| $z_j - c_j$ | | 0 | 0.1 | 0.1 | 0.1 | 0.101 | |

$s_1$, $s_2$ and $s_3$ are the *surplus* variables of the first three constraints, and $s_4$, $s_5$ and $s_6$ are the *slack* variables of the last three constraints.

The nondominated solution obtained for the original problem is $\mathbf{x} = (3.25, 1.15, 0.6)$ and the corresponding image in the objective function space is $\mathbf{z} = (3.25, 1.15, 0.6)$.

The variation intervals of the objective functions are updated: $[LB_1, UB_1] = [3.25, 7]$, $[LB_2, UB_2] = [1.15, 6]$, $[LB_3, UB_3] = [0.6, 6]$. These intervals are updated whenever a new solution is computed in order that the objective function values belong to the corresponding interval.

The range of variation of $t$ keeping the optimal basis, when $\mathbf{b}$ changes to $\mathbf{b} + t\ \mathbf{d}$, is computed through sensitivity analysis. If $\mathbf{x}_B$ is the basic variable vector, then, for $t = 0$, $\mathbf{x}_B = \mathbf{B}^{-1}\ \mathbf{b}$, where $\mathbf{B}^{-1}$ is the inverse of the basis matrix. Changing $\mathbf{b}$ to $\mathbf{b} + t\ \mathbf{d}$, $\mathbf{x}_B(t) = \mathbf{B}^{-1}\ (\mathbf{b} + t\ \mathbf{d})$, with $t$ taking values that guarantee the basis remains feasible, i.e. $\mathbf{x}_B(t) \geq \mathbf{0}$.

$$\mathbf{x}_B = \begin{bmatrix} x_3 \\ s_5 \\ s_6 \\ x_1 \\ x_2 \\ v^+ \end{bmatrix} = \mathbf{B}^{-1}\ (\mathbf{b} + t\ \mathbf{d})$$

$$= \begin{bmatrix} -0.4 & -0.4 & 0.6 & 0.4 & 0 & 0 \\ 0.7 & -1.3 & 0.7 & -1.7 & 1 & 0 \\ -0.85 & -1.85 & 2.15 & -2.15 & 0 & 1 \\ 0.75 & -0.25 & -0.25 & 0.25 & 0 & 0 \\ -0.35 & 0.65 & -0.35 & 0.35 & 0 & 0 \\ 0.1 & 0.1 & 0.1 & -0.1 & 0 & 0 \end{bmatrix} \left( \begin{bmatrix} 6 \\ 5 \\ 5 \\ 5 \\ 9 \\ 16 \end{bmatrix} + t \begin{bmatrix} 2.5 \\ 3.5 \\ 4 \\ 0 \\ 0 \\ 0 \end{bmatrix} \right)$$

$$= \begin{bmatrix} 0.6 \\ 1.7 \\ 1.65 \\ 3.25 \\ 1.15 \\ 1.1 + t \end{bmatrix} \geq \begin{bmatrix} 0 \\ 0 \\ 0 \\ 0 \\ 0 \\ 0 \end{bmatrix}$$

Therefore, $t \in [-1.1, +\infty]$. Since currently $t = 0$, the lower variation is $t_1 = 1.1$ and the upper bound of variation is $t_2 = +\infty$, that is, $[t - t_1, t + t_2] = [-1.1, +\infty]$.

Suppose that the DM wishes to continue the search along a trajectory defined by this reference direction, and that he/she chooses the option to **proceed**. However, since $t_2 = +\infty$, it is not possible to continue the search in this direction because any point of the unbounded linear segment $\mathbf{b} + t\ \mathbf{d}$ would be projected onto the current solution $\mathbf{z} = \mathbf{x} = (3.25, 1.15, 0.6)$. Hence, the DM is asked to specify the objective function that he/she wishes to improve.

**Mandatory Change of the Direction**

Suppose that the DM wants to improve $f_3(\mathbf{x})$. Then, $d_3$ is changed to $d_3 + \sigma(UB_3 - LB_3)$, where $\sigma$ is a scalar, for example, equal to 0.5 (value suggested by the authors of the method):

$$d_1 = 2.5;\ \ d_2 = 3.5;\ \ d_3 = 4 + 0.5\ \ (6 - 0.6) = 6.7;$$

The vector $\mathbf{d}$ is normalized making $\sum_{j \in G} d_j = s = 10$ :

$$d_j \leftarrow d_j \frac{10}{2.5 + 3.5 + 6.7}, \quad j = 1, \ldots, 3$$

The new direction is $\mathbf{d} = \begin{bmatrix} 1.969 \\ 2.756 \\ 5.276 \\ 0 \\ 0 \\ 0 \end{bmatrix}$

The weights are also updated by decreasing $w_3$, and then the vector is normalized following a similar procedure as for **d**. For example, a possible way is dividing $w_3$ by $(1+\sigma)$:

$$w_1 = 2.5; \;\; w_2 = 3.5; \;\; w_3 = \frac{4}{1.5} = 2.667;$$

After normalization $\mathbf{w} = \begin{bmatrix} 2.885 \\ 4.038 \\ 3.079 \\ 0 \\ 0 \\ 0 \end{bmatrix}$

Note that the weights **w** produce an effect on the objective function values that is, in general, the opposite of what might be expected; that is, a lower weight means a higher importance assigned to the corresponding objective function. This occurs when the reference point is located on the frontier or outside the feasible region, that is, when $v^+ \geq 0$ and $v^- = 0$.

The reference point is updated with the objective function values of the last solution[4]:

$$\mathbf{b}_G = \begin{bmatrix} 3.25 \\ 1.15 \\ 0.6 \end{bmatrix}$$

The following problem is solved considering $t = 0$:

$$\begin{aligned} \min \quad & v^+ - v^- - 0.001\ (x_1 + x_2 + x_3) \\ \text{s.t.} \quad & 2.885\ v^+ - 2.885\ v^- + x_1 \geq 3.25 + 1.969t \\ & 4.038\ v^+ - 4.038\ v^- + x_2 \geq 1.15 + 2.756t \\ & 3.079\ v^+ - 3.079\ v^- + x_3 \geq 0.6 + 5.276t \\ & x_1 + x_2 + x_3 \leq 5 \\ & x_1 + 3x_2 + x_3 \leq 9 \\ & 3x_1 + 4x_2 \leq 16 \\ & x_1, x_2, x_3, v^+, v^- \geq 0 \end{aligned}$$

[4] These weight and reference point updates were not proposed in the original presentation of the method. However, experimental results have shown that the performance of the method with these modifications is superior to the original version, especially concerning the changing direction control.

The optimal solution for $t=0$ is $\mathbf{x}=\mathbf{z}=(3.25,\ 1.15,\ 0.6)$ as was expected because this the reference point and it is reachable.

The range of variation of $t$ is computed regarding (4.4) such that the optimal basis is maintained:

$$\mathbf{x}_B(t) = \mathbf{B}^{-1}\ (\mathbf{b} + t\ \mathbf{d}) \geq \mathbf{0}$$

$$\begin{bmatrix} x_3 \\ s_5 \\ s_6 \\ x_1 \\ x_2 \\ v^+ \end{bmatrix} = \begin{bmatrix} -0.308 & -0.308 & 0.692 & 0.308 & 0 & 0 \\ 0.808 & -1.192 & 0.808 & -1.808 & 1 & 0 \\ -0.519 & -1.519 & 2.481 & -2.481 & 0 & 1 \\ 0.712 & -0.288 & -0.288 & 0.288 & 0 & 0 \\ -0.404 & 0.596 & -0.404 & 0.404 & 0 & 0 \\ 0.1 & 0.1 & 0.1 & -0.1 & 0 & 0 \end{bmatrix} \times$$

$$\left( \begin{bmatrix} 3.25 \\ 1.15 \\ 0.6 \\ 5 \\ 9 \\ 16 \end{bmatrix} + t \begin{bmatrix} 1.969 \\ 2.756 \\ 5.276 \\ 0 \\ 0 \\ 0 \end{bmatrix} \right) = \begin{bmatrix} 0.6 + 2.199\ t \\ 1.7 + 2.565\ t \\ 1.65 + 7.879\ t \\ 3.25 - 0.916\ t \\ 1.15 - 1.283\ t \\ 0 + t \end{bmatrix} \geq \begin{bmatrix} 0 \\ 0 \\ 0 \\ 0 \\ 0 \\ 0 \end{bmatrix}$$

meaning that $t \in [0,\ \ 0.897]$. Since currently $t=0$, then $t_1 = 0$ and $t_2 = 0.897$.

The DM can proceed with the search for solutions in this direction with the displacement being made with the basic speed ($\Delta t = \beta = 10^{-4}$—predefined scalar) or at a higher speed. An increase in speed means a larger separation between the nondominated solutions computed following the chosen trajectory, to be presented to the DM. So, the trajectory is travelled faster.

Suppose that the DM increases the speed to $\Delta t = 0.02$.

The trajectory presented to the DM is formed by the solutions to (4.4) for $t$ values spaced of $\Delta t$. These solutions, in particular, for $t=0.02$, are computed as follows ($x$ values in bold):

$$\begin{bmatrix} x_3 \\ s_5 \\ s_6 \\ x_1 \\ x_2 \\ v^+ \end{bmatrix} = \begin{bmatrix} 0.6 + 2.199 \times 0.02 \\ 1.7 + 2.565 \times 0.02 \\ 1.65 + 7.879 \times 0.02 \\ 3.25 - 0.916 \times 0.02 \\ 1.15 - 1.283 \times 0.02 \\ 0 + 0.02 \end{bmatrix} = \begin{bmatrix} \mathbf{0.644} \\ 1.751 \\ 1.808 \\ \mathbf{3.232} \\ \mathbf{1.124} \\ 0.02 \end{bmatrix}$$

Hence, **proceeding** in this direction (Fig. 4.37) the DM is faced, in a sequential and dynamical way, with a set of solutions:

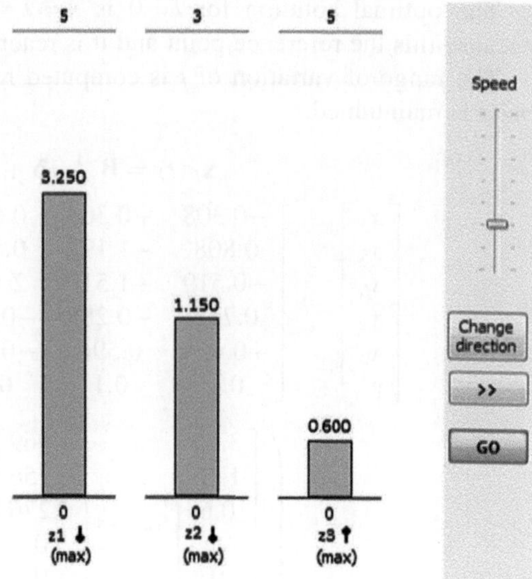


**Fig. 4.37** Bars of objective function values varying dynamically along the trajectory

| | $z_1$ | $z_2$ | $z_3$ |
|---|---|---|---|
| (t = 0.02) | 3.232 | 1.124 | 0.644 |
| (t = 0.04) | 3.213 | 1.099 | 0.688 |
| (t = 0.06) | 3.195 | 1.073 | 0.732 |
| (t = 0.08) | 3.177 | 1.047 | 0.776 |
| ... | ... | ... | ... |
| (t = 0.897) | 2.428 | 0 | 2.572 |

where $f_3(\mathbf{x})$ increases and $f_1(\mathbf{x})$ and $f_2(\mathbf{x})$ decrease.

When $t$ reaches the value 0.897 (solution where $z_1 = 2.428$, $z_2 = 0$ and $z_3 = 2.572$) an edge of the nondominated region is reached and so, in order to continue the search following the same direction, it is necessary to change the basis (see Fig. 4.40).

**Changing the Basis (Deciding to Keep the Search Direction)**

After updating the simplex tableau to change the basis (for example, using the dual simplex method), a new variation interval is computed for $t$. The result is $t_1 = 0$ and $t_2 = 1.580$, meaning that $t$ (currently equal to 0.897) can change within the interval $[0.897, 0.897 + 1.580]$.

Admitting that the DM wishes a slight increase in the speed (choosing $\Delta t = 0.03$), the following sequence of solutions is presented, where $f_3(\mathbf{x})$ keeps increasing, $f_1(\mathbf{x})$ is decreasing, and $f_2(\mathbf{x})$ stays constant (see Fig. 4.38):

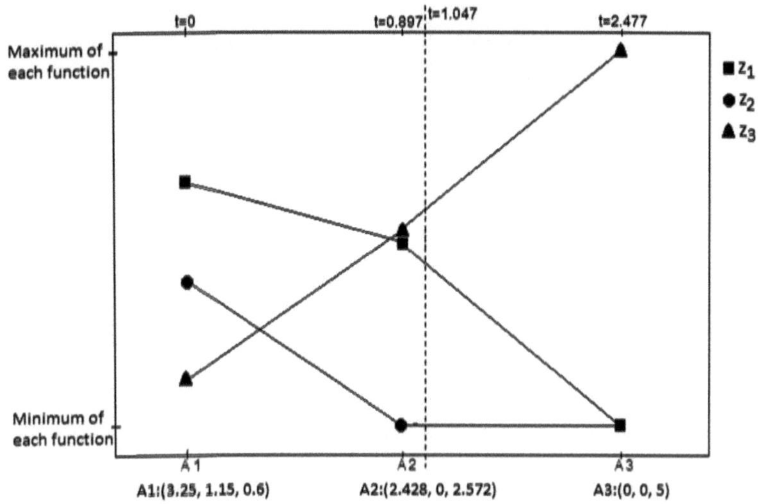


**Fig. 4.38** Objective function values along the first trajectory. The *dashed line* indicates the moment the DM changed the direction of search

| | $z_1$ | $z_2$ | $z_3$ |
|---|---|---|---|
| (t = 0.927) | 2.382 | 0 | 2.618 |
| (t = 0.957) | 2.335 | 0 | 2.655 |
| ... | ... | ... | ... |
| (t = 1.047) | 2.194 | 0 | 2.806 |

Suppose that, at this stage, analyzing the solution $\mathbf{z} = (2.194, 0, 2.806)$, the DM opts for changing the direction (before knowing the entire nondominated trajectory that could be computed following the current reference direction).

Figure 4.38 shows the values of $z_1$, $z_2$ and $z_3$, along the trajectory, and the dashed line ($t$ =1.047) indicates the moment in which the DM changed the direction of search.

**Voluntary Change of Direction**

Admitting that the DM wants to improve $f_2(\mathbf{x})$, the component corresponding to this objective function is increased in the direction **d** and the corresponding component of the weight vector **w** is decreased:

$$d_2 \leftarrow d_2 + \sigma\ (UB_2 - LB_2) = 2.756 + 0.5\ (6 - 0) = 5.756$$
$$w_2 \leftarrow \frac{w_2}{1.5}$$

After normalization:

$$\mathbf{d} = \begin{bmatrix} 1.514 \\ 4.428 \\ 4.058 \\ 0 \\ 0 \\ 0 \end{bmatrix} \quad \text{and} \quad \mathbf{w} = \begin{bmatrix} 3.333 \\ 3.111 \\ 3.555 \\ 0 \\ 0 \\ 0 \end{bmatrix}$$

The reference point is updated with the objective function values of the last solution:

$$\mathbf{b}_G = \begin{bmatrix} 2.194 \\ 0 \\ 2.806 \end{bmatrix}$$

The following problem is solved considering $t = 0$ (this problem corresponds to the problem of the first trajectory with $t = 1.047$):

$$\begin{array}{ll} \min & v^+ - v^- - 0.001(x_1 + x_2 + x_3) \\ \text{s.t.} & 3.333\ v^+ - 3.333\ v^- + x_1 \geq 2.194 + 1.514\ t \\ & 3.111\ v^+ - 3.111\ v^- + x_2 \geq 0 + 4.428\ t \\ & 3.555\ v^+ - 3.555\ v^- + x_3 \geq 0.806 + 4.058\ t \\ & x_1 + x_2 + x_3 \leq 5 \\ & x_1 + 3x_2 + x_3 \leq 9 \\ & 3x_1 + 4x_2 \leq 16 \\ & x_1, x_2, x_3,\ v^+, v^- \geq 0 \end{array}$$

After the sensitivity analysis of the right-hand side of the constraints it is concluded that the admissible interval for $t$ is [0, 1.206].

If the DM chooses again the option **proceed**, maintaining the previous speed ($\Delta t = 0.03$), a second trajectory is presented where $f_1(\mathbf{x})$ decreases, $f_2(\mathbf{x})$ increases and $f_3(\mathbf{x})$ increases:

| | $z_1$ | $z_2$ | $z_3$ |
|---|---|---|---|
| $(t = 0)$ | 2.194 | 0 | 2.806 |
| $(t = 0.03)$ | 2.138 | 0.04 | 2.821 |
| $(t = 0.06)$ | 2.083 | 0.121 | 2.852 |
| ... | ... | ... | ... |
| $(t = 0.42)$ | 1.417 | 0.562 | 3.020 |

Suppose that the DM considers the solution $\mathbf{z} = \mathbf{x} = (1.417, 0.562, 3.020)$ a good compromise solution. Then, the procedure stops.

Figure 4.39 shows the values of $z_1$, $z_2$ and $z_3$ along the entire trajectory and the moment when the DM decided to stop the search, represented by a dashed line.

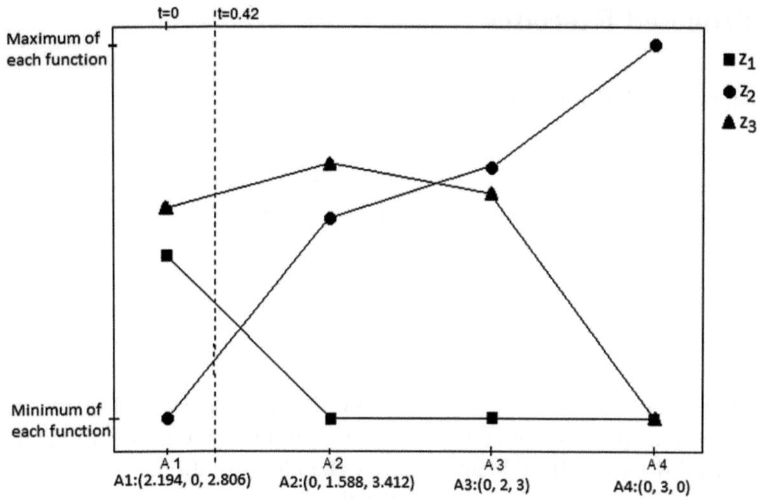


**Fig. 4.39** Objective function values along the second trajectory. The *dashed line* indicates the moment when the DM decided to stop the search

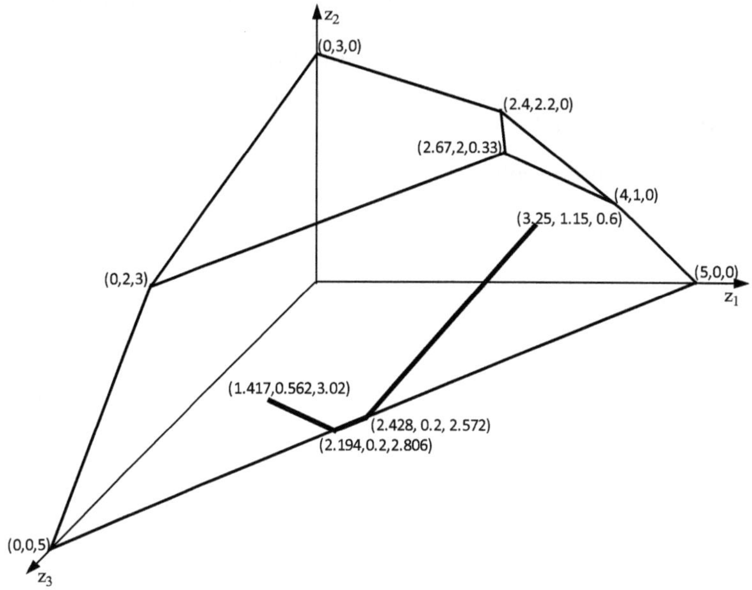


**Fig. 4.40** Trajectory on the nondominated frontier

Figure 4.40 presents the objective function space, which in this example coincides with the decision space, where the set of examined nondominated solutions is represented using a solid thick line.

## 4.7 Proposed Exercises

1. Consider the MOLP problem:

$$\begin{aligned} \max \quad & z_1 = f_1(\mathbf{x}) = -x_1 + 4x_2 \\ \max \quad & z_2 = f_2(\mathbf{x}) = 3x_1 - x_2 \\ \text{s.t.} \quad & -x_1 + x_2 \le 6 \\ & x_1 + x_2 \le 10 \\ & x_1 \le 8 \\ & x_2 \le 7 \\ & x_1, x_2 \ge 0 \end{aligned}$$

   (a) Represent the feasible region in the decision space and in the objective function space. Identify the efficient region and the nondominated region.
   (b) Build the pay-off table and identify the ideal solution.
   (c) Formulate the problem to find the first compromise solution according to the STEM method.
   (d) Considering that the solution of the problem formulated in (c) is $(x_1, x_2) = (5.46, 4.54)$, formulate the problem to determine the second solution generated by the STEM method if the DM decides relaxing $f_1(\mathbf{x})$ by 2 units. Graphically illustrate the reduction of the feasible region in the decision space and in the objective function space.
   (e) What is the new compromise solution obtained by solving the problem formulated in (d)?

2. Consider the MOLP problem:

$$\begin{aligned} \min \quad & z_1 = f_1(\mathbf{x}) = x_1 \\ \max \quad & z_2 = f_2(\mathbf{x}) = x_2 \\ \text{s.t.} \quad & \mathbf{x} \in X \end{aligned}$$

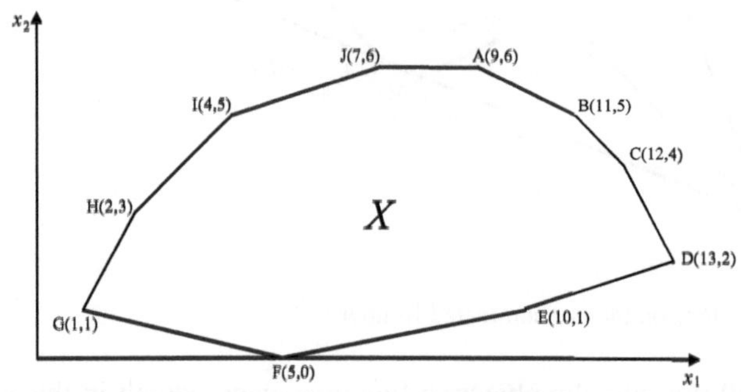


   (a) Formulate the problem to solve the initial iteration of the STEM method.
   (b) Compute the first solution proposed to the DM.

(c) Suppose that a second iteration is performed and the DM decides to relax $f_1(\mathbf{x})$ by 2 units with respect to the solution obtained in (b). Formulate the new problem to be solved and graphically illustrate the new feasible region.
(d) What is the second solution proposed by the STEM method?

3. Consider the following MOLP problem:

$$\begin{array}{ll} \max & z_1 = f_1(\mathbf{x}) = x_1 + 5x_2 \\ \max & z_2 = f_2(\mathbf{x}) = 6x_1 + 2x_2 \\ \text{s.t.} & x_1 + x_2 \leq 10 \\ & 3x_1 + x_2 \leq 24 \\ & x_1, x_2 \geq 0 \end{array}$$

(a) Formulate the problem to be solved in the first iteration of the STEM method.
(b) Would it be possible to reach a weakly efficient solution in the STEM method? If so, indicate a solution in these circumstances.

4. Consider the MOLP problem:

$$\begin{array}{ll} \max & z_1 = f_1(\mathbf{x}) = 3x_1 + x_2 + 2x_3 + x_4 \\ \max & z_2 = f_2(\mathbf{x}) = x_1 - x_2 + 2x_3 + 4x_4 \\ \max & z_3 = f_3(\mathbf{x}) = -x_1 + 5x_2 + x_3 + 2x_4 \\ \text{s.t.} & 2x_1 + x_2 + 4x_3 + 3x_4 \leq 60 \\ & 3x_1 + 4x_2 + x_3 + 2x_4 \leq 60 \\ & x_1 + 2x_2 + 3x_3 + 4x_4 \leq 50 \\ & x_1, x_2, x_3, x_4 \geq 0 \end{array}$$

(a) Find the indifference region on the weight space (use the projection on the plane $\lambda_1, \lambda_2$), corresponding to the nondominated basic solution that optimizes

$$\begin{array}{ll} \max & 0.1f_1(\mathbf{x}) + 0.6f_2(\mathbf{x}) + 0.3f_3(\mathbf{x}) \\ \text{s.t}: & \mathbf{x} \in X \quad (X \text{ denotes the feasible region defined above}) \end{array}$$

Comment on the solution obtained.
(b) What are the efficient nonbasic variables for this solution? What is the variation of each objective function per unit of each efficient nonbasic variable that becomes basic?
(c) What are the nondominated basic solutions that are obtained when each of these efficient nonbasic variables becomes basic? Graphically illustrate the corresponding indifference regions.
(d) After contracting the gradient cone (as in the ICW method) around the vector of central weights ($^1/_3$, $^1/_3$, $^1/_3$), what are the solutions found in (a) and (c) still possible to reach? Make the graphical analysis of the weight space.

(e) What are the constraints being introduced into the weight space if the DM prefers all the adjacent nondominated vertices regarding the solution computed in (a) in the method of Zionts and Wallenius?

(f) Comment on the following statement: "The nondominated solutions already known are not sufficient to completely characterize a nondominated face".

(g) Let $\mathbf{x}^0 = (5.73, 7.16, 1.63, 6.26)$ be the initial nondominated solution obtained by the STEM method. Formulate the problem to be solved in the next iteration of the STEM method if the DM decides to relax $f_3(\mathbf{x})$ by 14 units, with respect to that solution.

(h) Consider the auxiliary problem:

$$\begin{array}{ll} \min & \mathrm{v} \\ \text{s.t.} & \mathbf{x} \in X \\ & 10 - f_1(\mathbf{x}) \leq \mathrm{v} \\ & 40 - f_2(\mathbf{x}) \leq \mathrm{v} \\ & 20 - f_3(\mathbf{x}) \leq \mathrm{v} \\ & \mathrm{v} \geq 0 \end{array}$$

Is the solution to this auxiliary problem a nondominated solution to the initial problem?

If not, what would be the changes required in the formulation of the auxiliary problem to guarantee a nondominated solution?

(i) Comment on the following statements, regarding the original multiobjective problem:

1. "All the nondominated (basic and nonbasic) solutions already known belong to the same nondominated face".
2. "The nondominated basic solutions already known are sufficient to characterize the plane where a nondominated face is located, but not the face itself".
3. "There are no nondominated solutions that are alternative optima of at least one objective function."
4. "It is possible to obtain a nondominated solution that optimizes the objective function $f_k(\mathbf{x})$ by optimizing a weighted-sum of the objective functions with $\lambda_k = 0$ for $k = 1, 2, 3$".

5. Consider a linear programming problem with 3 objective functions being maximized, 4 decision variables and 3 constraints of the type '≤' (where $s_1$, $s_2$ and $s_3$ are the corresponding *slack* variables). Suppose that the weight space decomposition is the following:

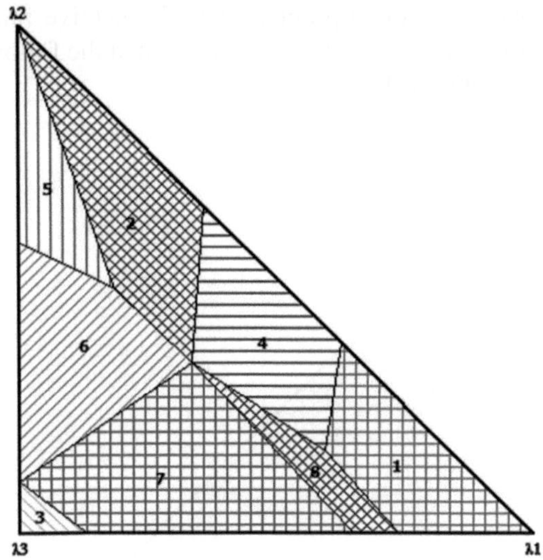


(a) Make the correspondence between each of the indifference regions 1–5 in the figure and the solutions A–E in the following table:

| | $z_1$ | $z_2$ | $z_3$ | Basic variables |
|---|---|---|---|---|
| A | 12.5 | 50 | 25 | $x_4=12.5$; $s_1=22.5$; $s_2=35$ |
| B | 15 | −15 | 75 | $x_2=15$; $s_1=45$; $s_3=20$ |
| C | 51 | 50 | 4 | $x_1=14$; $x_4=9$; $s_1=5$ |
| D | 66 | 30 | −12 | $x_1=18$; $x_3=6$; $s_3=14$ |
| E | 55.5 | 47.2 | 2 | $x_1=14.5$; $x_3=2.5$; $x_4=7$ |

(b) Identify the nondominated faces and edges of the problem.

6. Consider a MOLP with $p$ objective functions for which a first compromise solution is obtained through the resolution of the following linear problem:

$$\begin{aligned} \min \quad & \alpha \\ \text{s.t.} \quad & \lambda_k \left( z_k^* - \mathbf{c}_k \, \mathbf{x} \right) \leq \alpha \quad k = 1, \ldots, p \\ & \mathbf{Ax} = \mathbf{b} \\ & \mathbf{x} \geq \mathbf{0} \\ & \alpha \geq 0 \end{aligned}$$

where $z_k^*$ is the $k^{\text{th}}$ component of vector $\mathbf{z}^*$, which represents the ideal solution.

(a) Reformulate this problem to obtain a linear parametric programming problem that allows computing the nondominated solutions trajectory, when the reference point is displaced from the ideal solution in a direction $\mathbf{d}$.

(b) For simplicity reasons it has been assumed that the resolution of the problem above always leads to nondominated solutions. Is this truth?

7. Consider a linear programming problem with 3 objective functions, 4 decision variables and 2 constraints of the type '≤', for which the following weight space decomposition was obtained

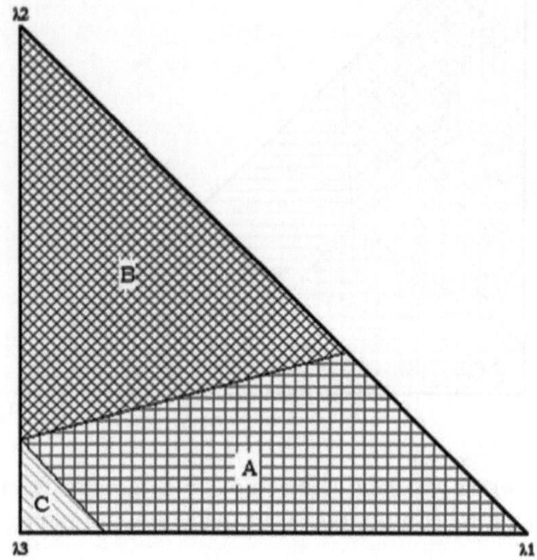


| | $z_1$ | $z_2$ | $z_3$ | Basic variables |
|---|---|---|---|---|
| A | 60 | 36 | 72 | $x_3 = 12$; $x_2 = 12$ |
| B | 45 | 63 | 66 | $x_4 = 18$; $x_2 = 6$ |
| C | 45 | 22.5 | 75 | $x_5 = 45$; $x_2 = 15$ |

(a) Identify the nondominated faces and edges of the problem.
(b) Suppose that first solution obtained using the method of Zionts-Wallenius is C, which is associated with the following simplex tableau:

| | $x_1$ | $x_2$ | $x_3$ | $x_4$ | $x_5$ | $x_6$ | |
|---|---|---|---|---|---|---|---|
| $x_5$ | 1.25 | 0 | 3.75 | 2.5 | 1 | −0.25 | 45 |
| $x_2$ | 0.75 | 1 | 0.25 | 0.5 | 0 | 0.25 | 15 |
| $z_1-c_1$ | 0.75 | 0 | −1.25 | 0 | 0 | 0.75 | |
| $z_2-c_2$ | 1.125 | 0 | −1.125 | −2.25 | 0 | 0.375 | |
| $z_3-c_3$ | 4.75 | 0 | 0.25 | 0.5 | 0 | 1.25 | |

$x_5$ and $x_6$ are the *slack* variables of the constraints.
Regarding the first iteration of the method, indicate:
1. the solution pairs that are presented to the DM for evaluation and the constraints resulting from the answers given by the DM.

2. The possible compromise vectors between the objectives (*trade-offs*) presented to the DM and the constraints resulting from the answers given by the DM.

8. Consider the following MOLP problem:

$$\begin{array}{ll} \max & z_1 = f_1(\mathbf{x}) = x_1 \\ \max & z_2 = f_2(\mathbf{x}) = x_2 \\ \max & z_3 = f_3(\mathbf{x}) = x_3 \\ \text{s.t.} & x_1 + x_2 + x_3 \le 5 \\ & x_1 + 3x_2 + x_3 \le 9 \\ & 3x_1 + 4x_2 \quad \le 16 \\ & x_1 + 2x_3 \quad \le 10 \\ & x_1, x_2, x_3 \ge 0 \end{array}$$

(a) This is a problem with degenerate solutions. How many bases correspond to the vertex that optimizes $f_3(\mathbf{x})$? Compute those basic solutions.
(b) Find the indifference regions on the weight space corresponding to each basic solution.

9. Consider the following MOLP problem:

$$\text{Max} \quad \mathbf{z} = \mathbf{f}(\mathbf{x}) = \begin{bmatrix} 3 & 1 & 2 & 1 \\ 1 & -1 & 2 & 4 \\ -1 & 5 & 1 & 2 \end{bmatrix} \begin{bmatrix} x_1 \\ x_2 \\ x_3 \\ x_4 \end{bmatrix}$$

$$\text{s.t.} \qquad \mathbf{Ax} \le \mathbf{b}, \quad \mathbf{x} \ge \mathbf{0}$$

where $\mathbf{A} = \begin{bmatrix} 2 & 1 & 4 & 3 \\ 3 & 4 & 1 & 2 \end{bmatrix}$ and $\mathbf{b} = \begin{bmatrix} 60 \\ 60 \end{bmatrix}$

Suppose that the following information is obtained using the TRIMAP method:

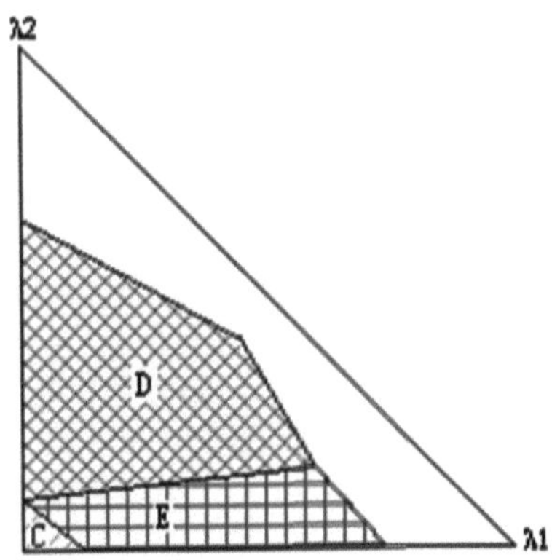


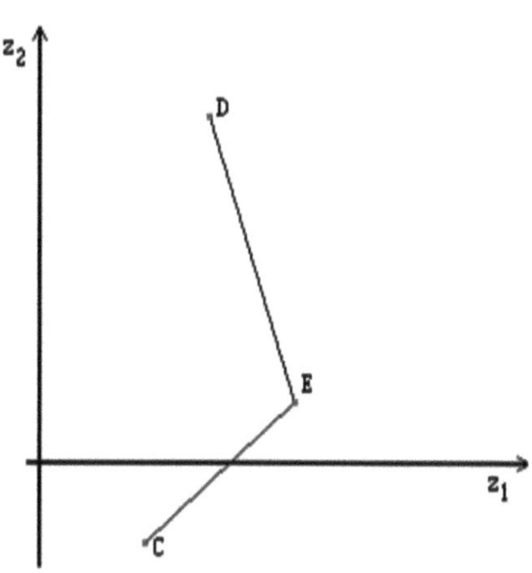

(a) What are the nondominated edges already known?
(b) Is there any feasible face completely known? Is it a nondominated face?
(c) Consider that the DM is only interested in continuing the search for nondominated solutions that satisfy $f_2(\mathbf{x}) \geq 28$. Find the region of the weight space to keep searching considering the constraint just introduced.

# Chapter 5
# A Guided Tour of iMOLPe

## 5.1 Introduction

The *interactive MOLP explorer* (iMOLPe) software is a computational package to deal with MOLP problems, which has been developed by the authors and accompanies this book. This computational package is mainly designed for teaching and decision support purposes in MOLP problems. The aim is to offer students in engineering, management, economics and applied mathematics an intuitive environment as the entrance door to multiobjective optimization in which the main theoretical and methodological concepts can be apprehended through experimentation, thus enabling them to learn at their own pace (Alves et al. 2015).

The iMOLPe software offers a user-friendly environment with graphical interface. The main characteristics and distinctive features of iMOLPe are: it includes different search strategies and solution computation techniques that can be freely used to explore the nondominated solution set of the problem; it integrates some structured interactive methods (currently, the STEM, Interval Criterion Weights—ICW and Pareto Race methods, as well as most features of the TRIMAP method); it provides several result displays and graphics that interconnect the information that is being gathered.

This chapter is not intended to be an extensive guide of the software because not all the available options are described herein. The reader can find a detailed description of each operation in the 'Help' menu of the software. This chapter aims to provide a guided tour of the software through examples, highlighting some of its most relevant features.



C. Henggeler Antunes et al., *Multiobjective Linear and Integer Programming*,
EURO Advanced Tutorials on Operational Research,
DOI 10.1007/978-3-319-28746-1_5

## 5.2 iMOLPe: Interactive MOLP Explorer

The iMOLPe software has been implemented in Delphi for Windows and uses the free code *lpsolve55* for solving the LP scalar problems. The main features included in iMOLPe are the following:

- Different scalarizing techniques for computing nondominated solutions (presented in Chap. 3), namely the optimization of weighted-sums of the objective functions, optimization of one of the objectives considering the others as constraints (*e-constraint* technique) and projection of a reference point onto the nondominated frontier—these techniques are available through the 'Compute' menu (Fig. 5.1). The weighted-sum and the reference point scalarizing techniques can be used alone or with additional constraints on the objective function values—option available in the 'Limits' menu (Fig. 5.1).
- Search strategies and visualization of results that are features of the TRIMAP method (presented in Chap. 4), which consists of a learning oriented set of tools dedicated to problems with three objective functions:
  - whenever a nondominated solution is computed by optimizing a weighted-sum of the objective functions, its indifference region on the weight space is also computed and graphically displayed—'Weight Space' window shown in Fig. 5.3 (rectangle) for a two-objective problem and in Fig. 5.8 (triangle) for a three-objective problem;
  - graphical display of the 2D or 3D objective function space, showing the nondominated points already computed and the nondominated edges connecting adjacent nondominated vertices—'Solution Graphs' shown, e.g., in Fig. 5.6 for a two-objective problem (2D) or in Fig. 5.7 for a three-objective problem (3D); this graph as well as the 'Weight Space' are shown by selecting the respective items in the 'Solutions' menu (Fig. 5.1);

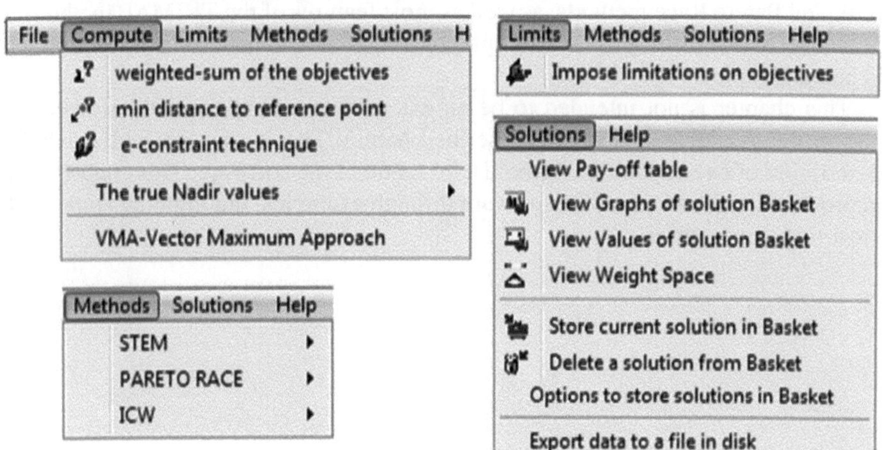


**Fig. 5.1** Main menu of the iMOLPe software

- conversion of bounds specified for the objective function values into the weight space, that is, computation and graphical representation of the region of the weight space defined by the weight vectors leading to (known and unknown) nondominated solutions satisfying those bounds; this feature is available through the 'Calculate' menu in the 'Weight Space' window (Fig. 5.9);

- There are other options (not originally included in TRIMAP) available on the graph windows mentioned above by means of local menus or buttons; for instance, the graphical representation on the weight space of constraints resulting from preferences expressed as *"solution x is preferred to solution y"*, as in the Zionts–Wallenius method (Fig. 5.9);
- Integration of the following interactive methods (presented in Chap. 4): STEM, Pareto Race and ICW—'Methods' menu (Fig. 5.1); each method has a submenu for its specific operations.
- Options to manage the nondominated solutions already computed, such as choosing which solutions are kept in memory (*solution basket*), saving solutions in disk or choosing different means for visualizing results—'Solutions' menu (Fig. 5.1).
- Implementation of an exact procedure to compute the nadir point (proposed in Alves and Costa 2009)—'Compute' menu (Fig. 5.1). The nadir point gives the minimum (worst) objective function values over the set of all nondominated solutions.
- Although the intention of the software is to be mainly used as an interactive explorer, it further includes a *VMA-vector maximum algorithm* (Steuer 1986) that computes all nondominated basic solutions (vertices) to the MOLP problem—'Compute' menu (Fig. 5.1). In problems with up to three objective functions, the corresponding indifference regions are displayed on the parametric weight diagram ('Weight Space') and the nondominated points and edges in the objective space are shown in the 'Solution Graphs'. This algorithm can also be used within the ICW interactive method after the contraction of the criterion cone to compute a subset of all nondominated basic solutions that are reachable considering the reduced criterion cone (see step 5 of ICW in Chap. 4). This option is available under the specific menu of the ICW method.

In the following section we illustrate the use of the software with four examples. The first example considers a problem with two objective functions (to be maximized), the next two problems have three objective functions (also to be maximized) and the last problem has four objective functions (two to be maximized and two to be minimized).

### 5.2.1 Example 1

Consider the following bi-objective problem with four decision variables and four constraints:

$$
\begin{aligned}
\max z_1 = f_1(\mathbf{x}) &= -6x_1 + 30x_2 + 6x_3 + 12x_4 \\
\max z_2 = f_2(\mathbf{x}) &= 30x_1 + 10x_2 + 20x_3 + 10x_4 \\
\text{s.t.} & \\
2x_1 + x_2 + 4x_3 + 3x_4 &\le 60 \\
3x_1 + 4x_2 + x_3 + 2x_4 &\le 60 \\
x_1 + 2x_2 + 3x_3 + 4x_4 &\le 50 \\
4x_1 + 3x_2 + 2x_3 + x_4 &\le 50 \\
x_1, x_2, x_3, x_4 &\ge 0
\end{aligned}
$$

After creating the problem (or opening the respective data file if the problem has already been created), the iMOLPe software starts by computing the nondominated solutions that individually optimize each objective function and constitute the *pay-off* table (window (a) in Fig. 5.2). Every solution computed is shown on the main window ((b) in Fig. 5.2), which displays bar graphs for the objective values and numerical information (values of the variables and the objective functions). Each point is also depicted on the 2D graph that represents the objective space ('Solution Graphs'—window (c) in Fig. 5.2). These two solutions have been obtained by optimizing weighted-sums of the objective functions, considering a weight close to 1 for the objective that is being optimized and close to 0 for the other. The corresponding indifference regions on the weight space are drawn ('Weight Space'—window (d) in Fig. 5.2); since the problem has two objective functions, the weight space is a line segment from ($\lambda_1 = 0, \lambda_2 = 1$) to ($\lambda_1 = 1, \lambda_2 = 0$). This line needs, however, some "thickness" so that the indifferent regions can be visualized; so, it assumes a rectangle shape.

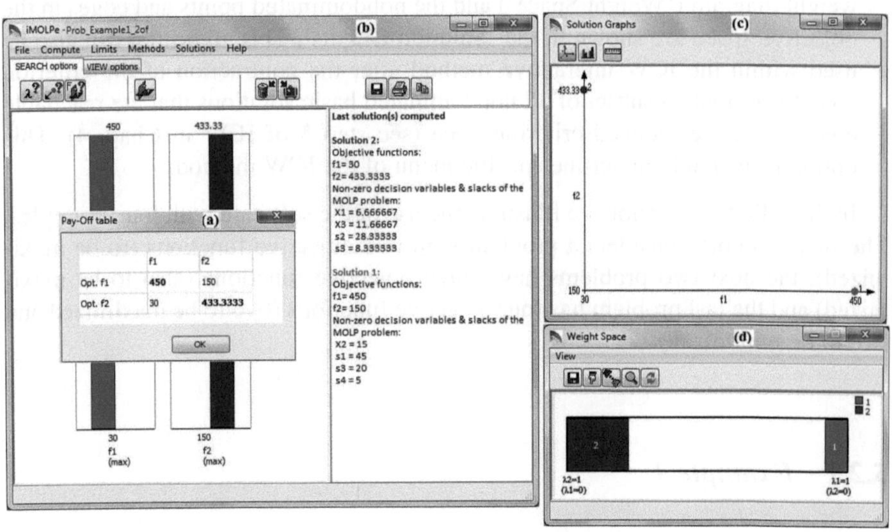


**Fig. 5.2** Example 1: nondominated solutions that individually optimize each objective function

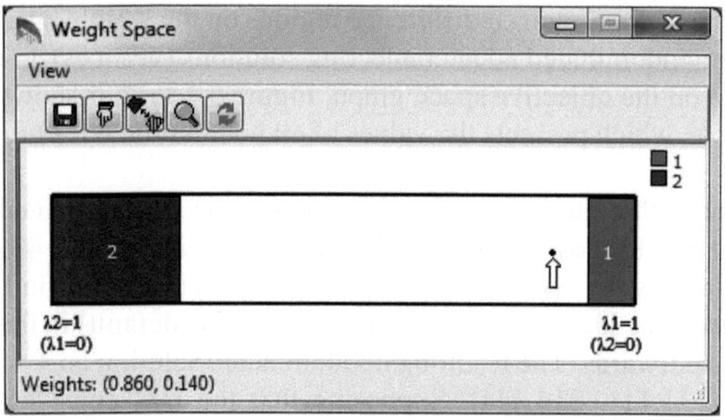


**Fig. 5.3** Example 1: the DM selects a new weight vector

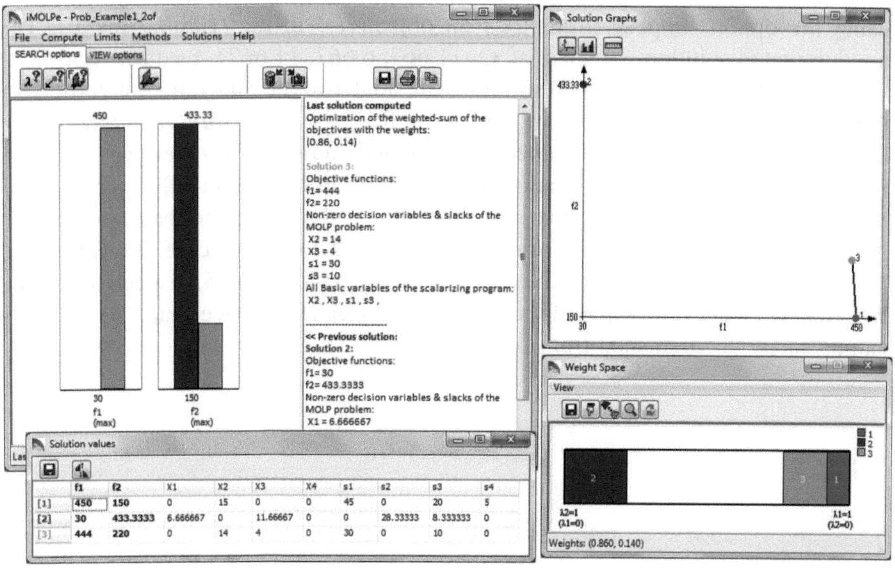


**Fig. 5.4** Example 1: optimization of a weighted-sum of the objective functions leading to solution 3

All the other tools to analyze the problem become now available. Some of them can be chosen either by means of a menu item or a button in the toolbar of the main window ((b) in Fig. 5.2), which has two tabs, one for 'SEARCH options' and another for 'VIEW options'.

Suppose that the DM chooses a new weight vector by clicking directly on a point within the blank area of the weight space to compute another nondominated basic solution. Consider that the weight vector pointed by the arrow in Fig. 5.3 is selected, i.e., $\boldsymbol{\lambda} = (0.86, 0.14)$.

The corresponding weighted-sum of the objective functions is optimized and solution 3 is obtained—see Fig. 5.4. Solutions 1 and 3 are found to be adjacent because they have contiguous indifference regions on the weight space. This means that there is a nondominated edge connecting solutions (vertices) 1 and 3, which is also displayed on the objective space graph. Figure 5.4 further shows the 'Solution values' window, which presents the values of all nondominated solutions computed thus far.

Now, suppose that the DM wishes to compute the nondominated solution that minimizes the Chebyshev distance to the ideal point, $\mathbf{z}^* = (450, 433.333)$. Therefore he/she chooses the option 'min distance to reference point' in the 'Compute' menu and selects the ideal point (which is presented by default) in the dialogue box that is shown afterwards. The resulting nondominated solution is: $\mathbf{x} = (0.481, 9.037, 10.481)$, $\mathbf{z} = (331.111, 314.444)$. Supposing that the DM considers the $z_2$ value unsatisfactory and wants to compute another solution close to the ideal point but imposing a lower bound of 350 on $f_2(\mathbf{x})$, then he/she calls again this computation process considering the additional limitation $f_2(\mathbf{x}) \geq 350$ (this limitation is included using the option 'impose limitations on objectives' in the 'Limits' menu). The nondominated solution obtained is: $\mathbf{x} = (1.667, 6.667, 11.667)$, $\mathbf{z} = (260, 350)$.

The two latter nondominated solutions are nonbasic (i.e., they are not vertices of the original nondominated region), which can be saved in the *solution basket* by the DM. Only basic solutions obtained by weighted-sums of the objective functions (i.e., vertices of the original nondominated region) are automatically saved, unless the option for saving all nondominated solutions—basic and nonbasic—has been previously indicated in the 'Solutions' menu (Fig. 5.1). In this case, the DM opted for saving these two solutions, thus they received the id. 4 and 5, respectively—see Fig. 5.5. They are represented by crosses on the objective space graph so that they can be easily distinguished from the vertices.

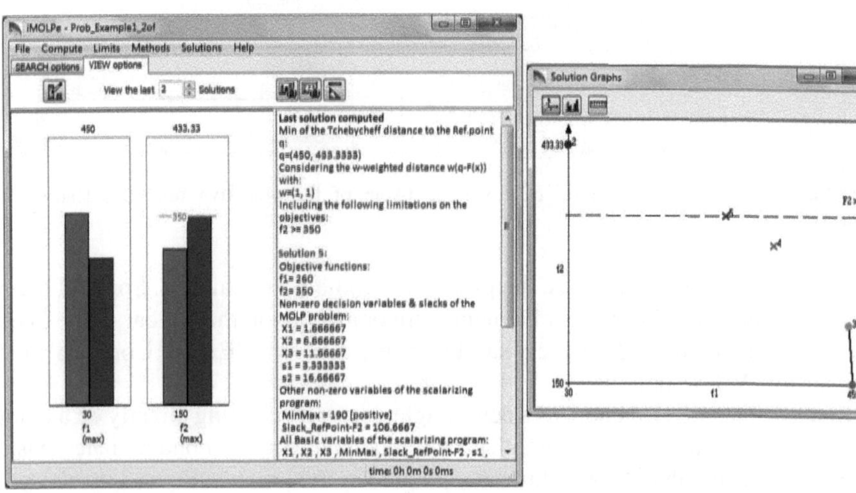


**Fig. 5.5** Example 1: minimizing the Chebyshev distance to the ideal point without additional limitations (solution 4) and with a lower bound $f_2(\mathbf{x}) \geq 350$ (solution 5)

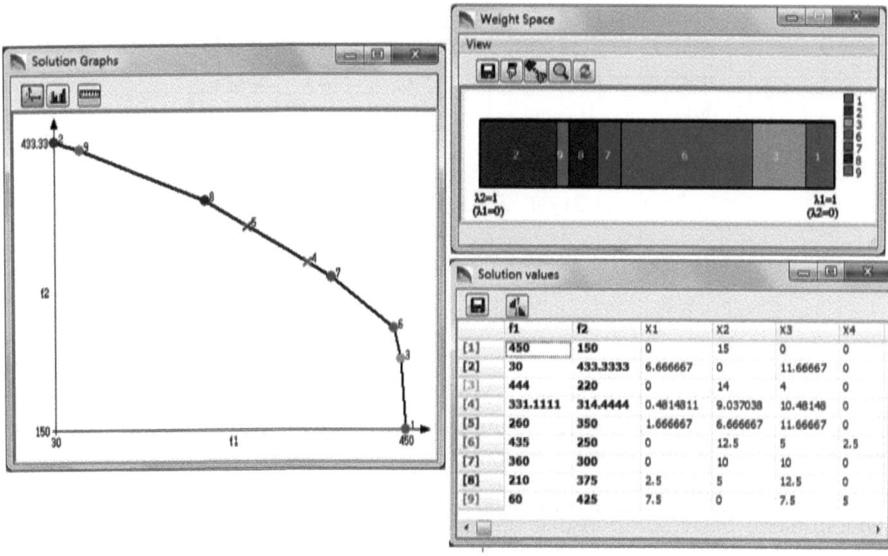


**Fig. 5.6** Example 1: all nondominated basic solutions plus two nondominated nonbasic solutions (4 and 5)

If the additional limitation on $f_2(\mathbf{x})$ is removed and all nondominated basic solutions are computed by optimizing weighted-sums of the objective functions (either by choosing weight vectors from unfilled areas of the weight space or by calling the *VMA-vector maximum algorithm*), then four additional nondominated solutions are obtained. This problem has seven nondominated vertices (solutions 1, 2, 3, 6, 7, 8 and 9 in Fig. 5.6). The nondominated edges (which are six) are also known, thus defining the whole nondominated set. Figure 5.6 shows the nondominated set on the objective function space, the decomposition of the weight space and the values of objective functions and variables of all solutions computed. The non-extreme points 4 and 5, previously computed using the reference point technique, are also displayed on the objective function space; they are located on the edge connecting vertices 7 and 8.

### 5.2.2 *Example 2*

Consider the following three-objective problem with the same feasible region as in Example 1 but different objective functions:

$$
\begin{array}{l}
\max z_1 = f_1(\mathbf{x}) = \ 18x_1 + 13x_2 + 12x_3 \\
\max z_2 = f_2(\mathbf{x}) = \ -6x_1 + 30x_2 + 6x_3 + 12x_4 \\
\max z_3 = f_3(\mathbf{x}) = \ 5x_1 - 6x_2 + 12x_3 + 24x_4 \\
\text{s.t.} \\
\qquad 2x_1 + x_2 + 4x_3 + 3x_4 \leq 60 \\
\qquad 3x_1 + 4x_2 + x_3 + 2x_4 \leq 60 \\
\qquad x_1 + 2x_2 + 3x_3 + 4x_4 \leq 50 \\
\qquad 4x_1 + 3x_2 + 2x_3 + x_4 \leq 50 \\
\qquad x_1, x_2, x_3, x_4 \geq 0
\end{array}
$$

As with the previous example, iMOLPe starts by computing the individual optima, i.e., the nondominated solutions that optimize each objective function $f_k(\mathbf{x})$, $k = 1, 2, 3$; these are the solutions 1, 2 and 3, respectively, in Fig. 5.7.

It can be recognized from the weight space that there will be at least one more nondominated basic solution that optimizes $f_1(\mathbf{x})$ because the indifference region of solution 1 does not completely fill the right corner of the triangle (where $\lambda = (1,0,0)$). This means that there will be other indifference region(s) including the point $\lambda_1 = 1$, $\lambda_2 = \lambda_3 = 0$. Let us suppose that the DM points toward the weight vector marked below region 1 on the weight space of Fig. 5.7. The optimization of the corresponding weighted-sum leads to solution 4 (Fig. 5.8). From the weight space it can be concluded that $f_1(\mathbf{x})$ has two (and only two) alternative optimal basic solutions. These are solutions 1 and 4, which trade-off the values of the other objective functions: $\mathbf{z}^1 = (260, 210, 132.5)$ and $\mathbf{z}^4 = (260, 30, 173.333)$.

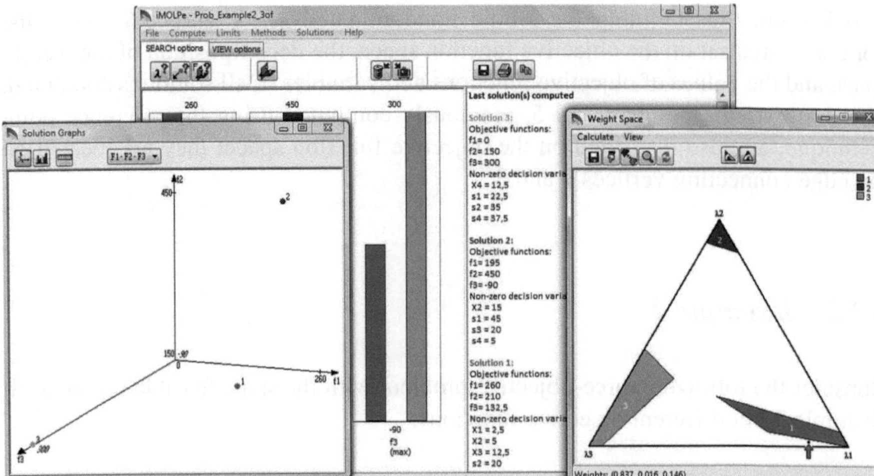


**Fig. 5.7** Example 2: first three nondominated solutions which individually optimize each objective function

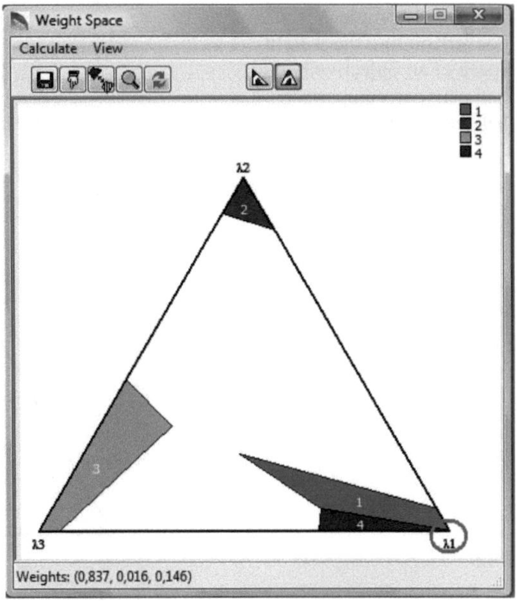


**Fig. 5.8** Example 2: solutions 1 and 4 are alternative optima for $f_1(\mathbf{x})$

As illustrated above in Example 1, the DM can include additional bounds on the objective function values (i.e., express inferior bounds below which solutions are not desired) and use them in combination with scalarizing techniques to compute new nondominated solutions in those restricted regions. However, this is not the only possibility of getting information resulting from imposing bounds on the objective functions. In problems with three objective functions, the DM can also visualize the graphical translation of those bounds into the weight space. This is a feature of the TRIMAP method that is illustrated in Fig. 5.9, which shows the region of the weight space corresponding to $f_2(\mathbf{x}) \geq 400$ (dashed area). Note that the selection of this feature (local option in the 'Weight Space' window) does not trigger the option 'Impose limitation on objectives' ('Limits' menu in the main window); it rather shows the weight area that leads to nondominated solutions that satisfy the bound(s) without the need of explicitly including them in the subsequent solution computation process.

The DM can also visualize the translation into the weight space of pairwise comparisons of nondominated solutions as in the Zionts–Wallenius method. The weight constraint resulting from the preference information "*solution 1 is preferred to solution 4*" is also displayed in Fig. 5.9.

A further option is available in the 'Calculate' menu of the 'Weight Space' window: it is to compute and show the area of the weight space where the nondominated solutions below the minimum already known for a given objective function are located, i.e., where the nadir value for that objective function can be found. The way this area is computed is described in Alves and Costa (2009). The nadir values are the minima attained by the objective functions over the set of all

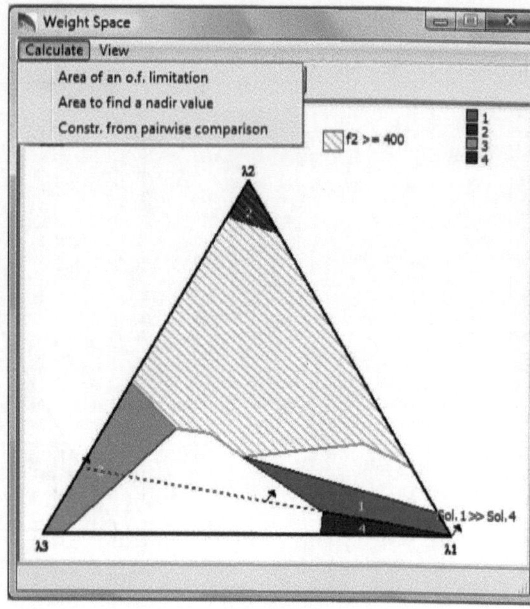


**Fig. 5.9** Example 2: conversion into the weight space of an objective bound and pairwise comparison between solutions

nondominated solutions. They provide valuable information for characterizing the ranges of the objective function values over this set. Together with the ideal values they define the nondominated bounds of each objective function, which are of major interest to allow a DM to size up the extent of variation of the objective function values in the region of interest. However, they are very difficult to determine except for the bi-objective case. The method proposed by Alves and Costa (2009) to compute the nadir values consists in the computation of the weight space region associated with the nondominated solutions that have a value below the minimum already known for the objective function under analysis. If this region is empty, the nadir value has been found; otherwise, a new nondominated solution is computed (through the optimization of a weighted-sum of the objective functions) using a weight vector picked from the delimited region. The process is repeated until the nadir value is found. The complete process to compute each nadir value for any number of objective functions is available in the 'Compute' menu—'The true nadir values' item. The option 'Area to find a nadir value' in the 'Calculate' menu of the weight space window only computes the weight space region where the nadir value of a specified objective function can be searched for (i.e., the first step of the method). This latter option is restricted to three objective functions as the region is graphically displayed.

Returning to Example 2, let us suppose that the DM chooses the interactive method ICW to continue the study of the problem. Firstly, the method optimizes weighted-sums of the objective functions using a pre-defined set of seven $(2p+1)$ weight combinations regularly dispersed in the weight space. These are the white dots indicated on the weight space of Fig. 5.10, three in the vertices of the triangle

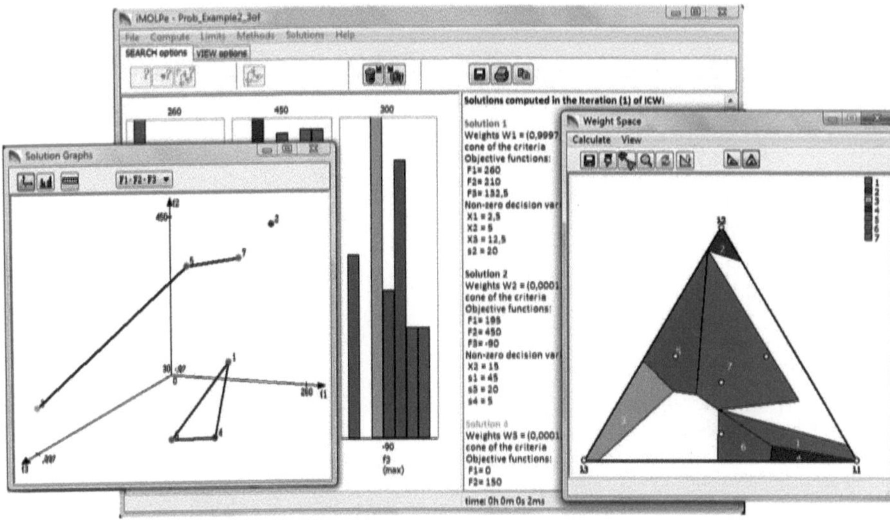


**Fig. 5.10** Example 2: first iteration of the ICW method

and four inside the triangle. The nondominated solutions 1, 2, 3, 5, 6 and 7 are obtained (note that solution 4 could have been obtained instead of solution 1). Solution 7 is computed twice, as two different weight combinations lead to this solution.

If a new iteration of the ICW method is performed, the DM is asked to choose his/her preferred solution from the previous sample of solutions. Suppose that the DM chooses solution 7. Since this solution was obtained for the sixth and seventh pre-defined weight combinations (see ICW method in Chap. 4), $\lambda^6 = (4/9, 4/9, 1/9)$ and $\lambda^7 = (1/3, 1/3, 1/3)$, the DM is further asked to select between these two weight vectors. Consider that the DM selects the most central one, $\lambda^7$. Then the criterion cone is contracted around the convex combination of the objectives given by this weight vector. The ICW method considers again the pre-defined set of weight combinations (white dots in the triangle on the right of Fig. 5.11), but now they are used for optimizing weighted-sums of the contracted objective functions. In Fig. 5.11 the graph on the right shows the weight space of the contracted criterion cone with the nondominated solutions obtained in this second iteration (solutions 5, 6, 7, 8 and 9), while the original weight space is shown on the left, presenting the nondominated solutions computed thus far; the region corresponding to the contracted criterion cone is also displayed inside the original weight space (dashed small triangle).

The nondominated region reachable with the contracted criterion cone can be completely searched using the option 'Contracted VMA' in the ICW menu (see Fig. 5.11). As a result, the nondominated basic solutions whose indifference regions fully decompose the weight space associated with the contracted criterion cone are computed (Fig. 5.12).

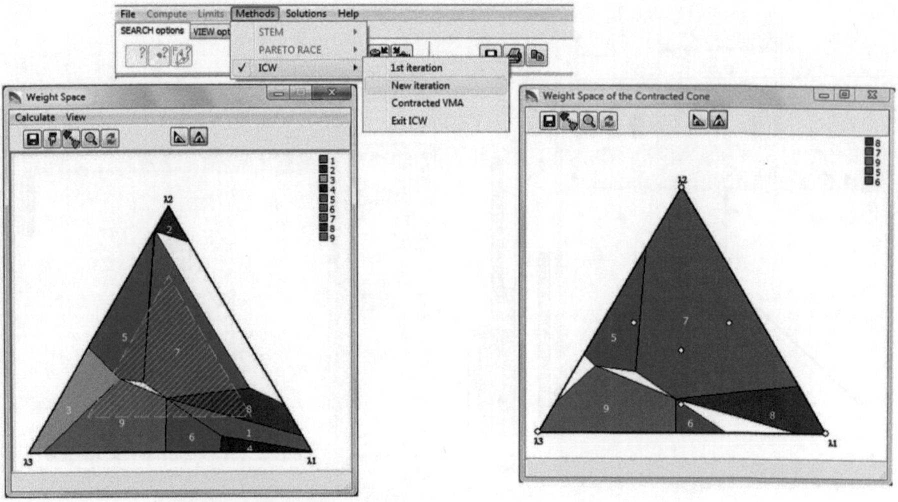


Fig. 5.11 Example 2: second iteration of the ICW method

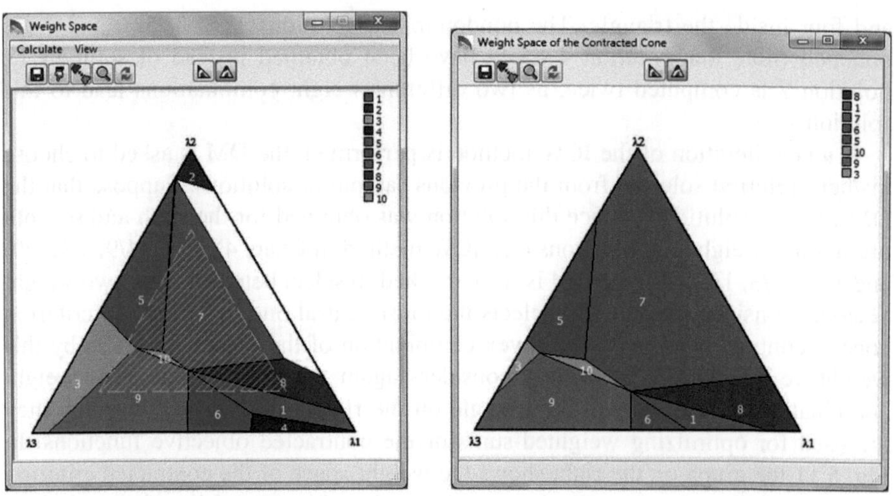


**Fig. 5.12** Example 2: computing all nondominated basic solutions that are reachable with the contracted cone of the objectives

As can be seen in the original weight space depicted on the left of Fig. 5.12, the nondominated solutions computed until now do not constitute the full set of nondominated basic solutions because there is still an unfilled area on the original weight space. If the DM wants to compute the whole nondominated solution set, he/she may exit the ICW method and ask for the optimization of weighted-sums of the objectives with weight vector(s) belonging to this unfilled area. One

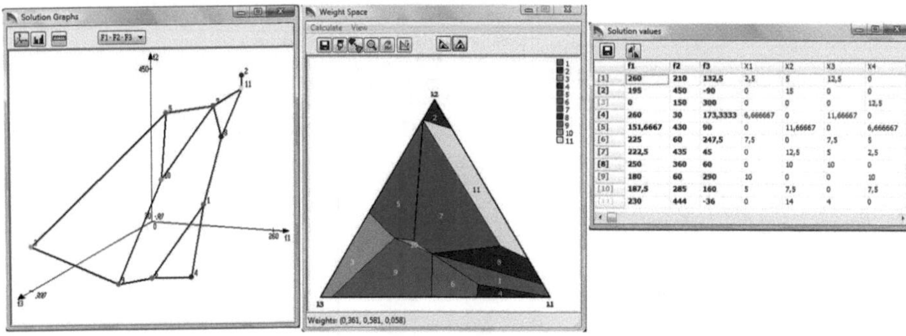


**Fig. 5.13** Example 2: all nondominated basic solutions of the problem are known

optimization is enough since there is only one more nondominated basic solution—solution 11. Figure 5.13 shows all solutions generated so far, including their numerical information. It can be observed that the problem has 11 nondominated vertices, 15 nondominated edges and 5 nondominated faces (which are defined by the vertices 1-4-6, 1-6-9-10-7-8, 3-5-10-9, 5-7-10 and 7-11-8). It is worth noting that points 2, 11, 7 and 5 are located on the same face but the solutions on the interior of the face are just weakly nondominated; they are dominated by the solutions on the edges 2-11, 11-7 and 7-5. A nondominated or weakly nondominated face is associated with a point of the weight space where three or more indifference regions intersect. However, while a (strictly) nondominated face is associated with a point in the interior of the weight space, a weakly nondominated face corresponds to a point in the border of the triangle, i.e., where one of the weights is equal to 0 ($\lambda_1 = 0$ for the face 2-11-7-5) – see Fig. 5.13.

Finally, suppose that the Pareto Race method is chosen. This method enables the DM to *travel* over the nondominated region as if he/she was driving a car, controlling the direction of motion (by selecting different objective functions to be improved) and the speed (obtaining solutions closer or farther from each other). Figure 5.14 shows the course that starts at the nondominated point obtained by the projection of the ideal point $\mathbf{z}^* = (260, 450, 300)$ and moves in a direction that improves $f_3(\mathbf{x})$.

### 5.2.3 *Example 3*

Now we provide a third example, also with three objective functions, in order to illustrate the decomposition of the weight space in a degenerate problem.

Consider the following problem:

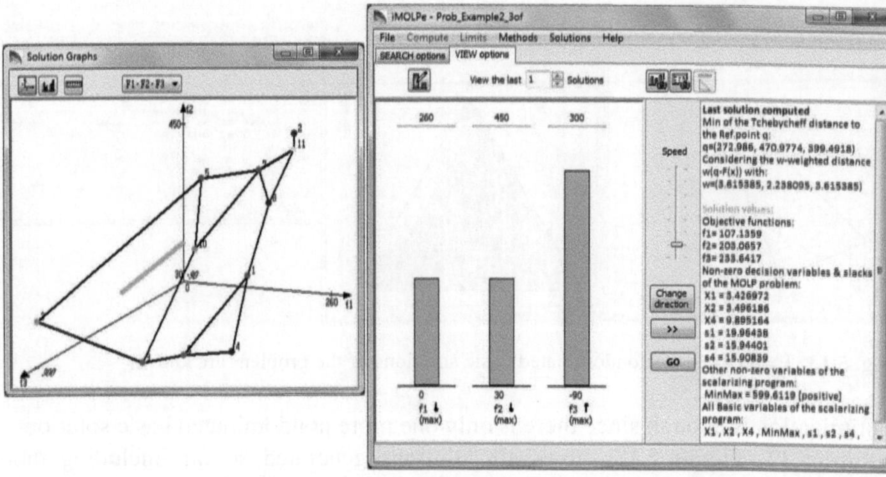


**Fig. 5.14** Example 2: traveling on the nondominated surface using the Pareto Race method

$$
\begin{aligned}
\max z_1 &= f_1(\mathbf{x}) = x_1 \\
\max z_2 &= f_2(\mathbf{x}) = x_2 \\
\max z_3 &= f_3(\mathbf{x}) = x_3 \\
\text{s.t.} & \\
& 5x_1 + 6x_2 + 3x_3 \leq 30 \\
& x_1 + x_2 + x_3 \leq 6 \\
& 5x_1 + 3x_2 + 6x_3 \leq 30 \\
& x_1, x_2, x_3 \geq 0
\end{aligned}
$$

At the stage shown in Fig. 5.15 all nondominated vertices of the problem have already been computed, although there are still unfilled areas on the weight space and not all the nondominated edges have been identified.

After completing the decomposition of the weight space the graph in Fig. 5.16 is obtained. As can be observed in this figure, solution 1 is degenerate and it is associated with three different bases. A nondegenerate basic solution corresponds to a single indifference region on the weight space. However, degenerate solutions have several bases, each one corresponding to an indifference region on the weight space. All nondominated edges leading to the degenerate solution can be identified as long as all bases have been computed.

The iMOLPe software acknowledges when different bases for the same solution are being computed. Thus, it keeps the same id. # of the solution (in this case, 1) and other bases are identified with #/2, #/3, etc. The indifference regions in Figs. 5.15 and 5.16 are represented using patterns instead of solid colors (an option available in the 'Weight Space' window) so that the different indifference regions associated with solution 1 can be better recognized. In general, indifference regions of degenerate solutions may exist that partially overlap each other. This is not the case of this example, as regions "1", "1/2" and "1/3" do not overlap. Exercise 5 in

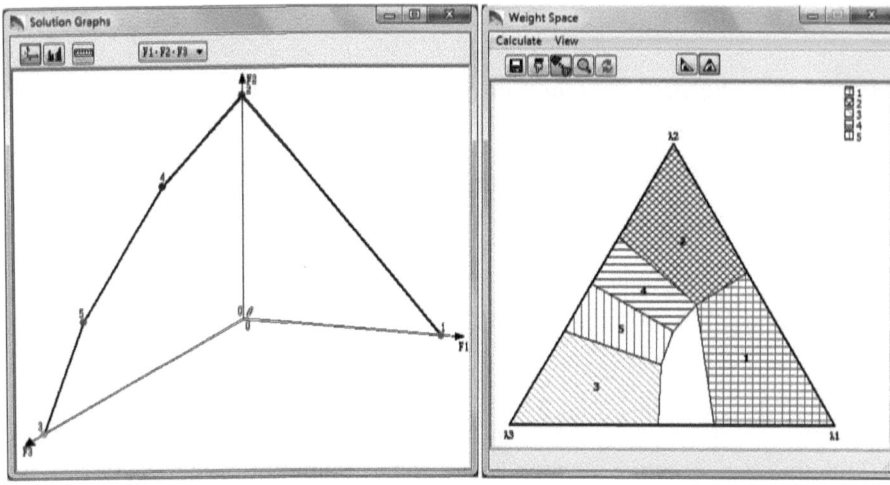


**Fig. 5.15** Example 3: all nondominated vertices are known but not all bases have been computed

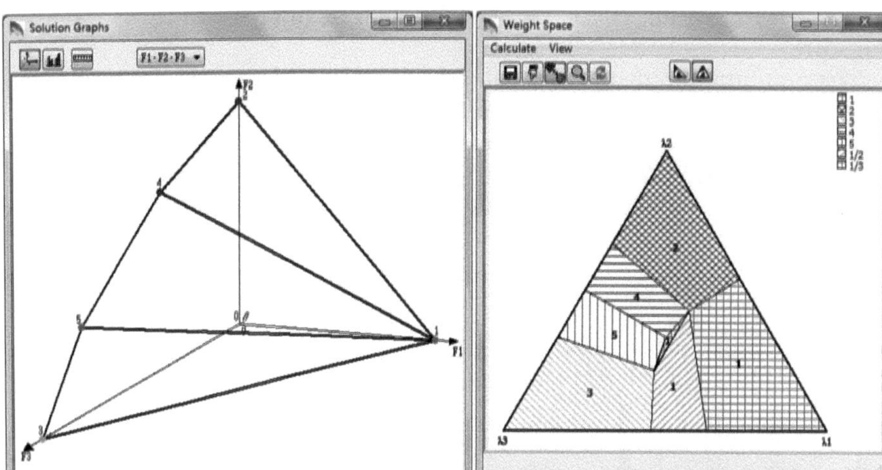


**Fig. 5.16** Example 3: several bases for the degenerate solution 1 lead to several indifference regions

Sect. 5.3 presents a problem in which there is overlapping of regions associated with different bases of a degenerate solution. The completion of this exercise is suggested to the reader to get a better understanding of this situation.

### 5.2.4 *Example 4*

Consider the following problem with four objective functions, two being maximized and two being minimized:

$$
\begin{array}{ll}
\max & z_1 = f_1(\mathbf{x}) = x_1 + 4x_2 + x_3 \\
\max & z_2 = f_2(\mathbf{x}) = 2x_1 + 3x_2 + 7x_3 \\
\min & z_3 = f_3(\mathbf{x}) = 10x_1 + 2x_2 + x_3 \\
\min & z_4 = f_4(\mathbf{x}) = x_1 + x_2 + 7x_3 \\
\text{s.t.} & x_1 + 2x_2 + x_3 \le 60 \\
 & 6x_1 + 3x_2 + 4x_3 \le 180 \\
 & x_1 + x_2 + x_3 \ge 10 \\
 & x_1, x_2, x_3 \ge 0
\end{array}
$$

Initially, each objective function is individually optimized and the *pay-off* table is built—see Fig. 5.17. Since the problem has more than three objective functions, neither the weight space nor the objective function space is shown.

Let us suppose that the DM wants to analyze this problem using the STEM method. Hence, he/she selects the 'first Iteration' item of the 'STEM' sub-menu (Fig. 5.18).

The nondominated solution computed is $\mathbf{x}^5 = (12.058,\ 16.823,\ 14.295)$, $\mathbf{z}^5 = (93.647,\ 174.650,\ 168.526,\ 128.945)$, which is shown in Fig. 5.19. The STEM method computes the nondominated solution closest to the ideal point $\mathbf{z}^* = (120, 315, 10, 10)$ according to a weighted Chebyshev metric (for details, see Chap. 4). The weights are automatically determined by the algorithm ($w$ in the

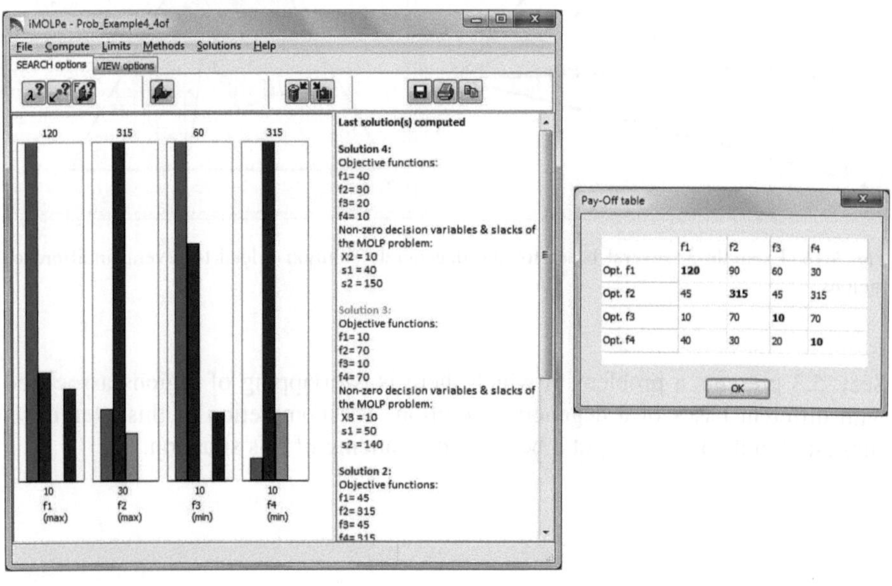


**Fig. 5.17** Example 4: nondominated solutions that individually optimize each objective function

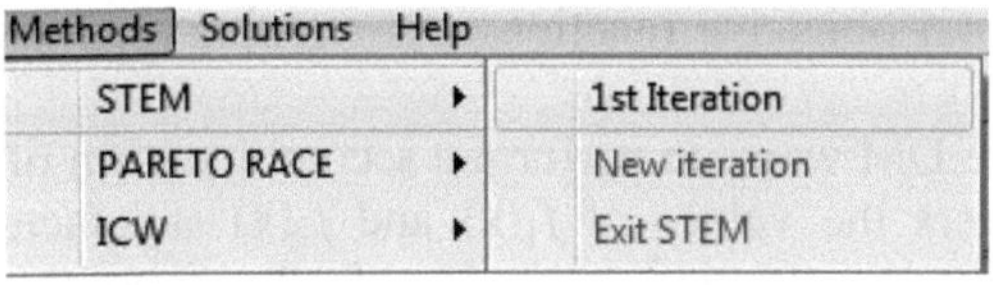


**Fig. 5.18** Example 4: choosing the STEM method

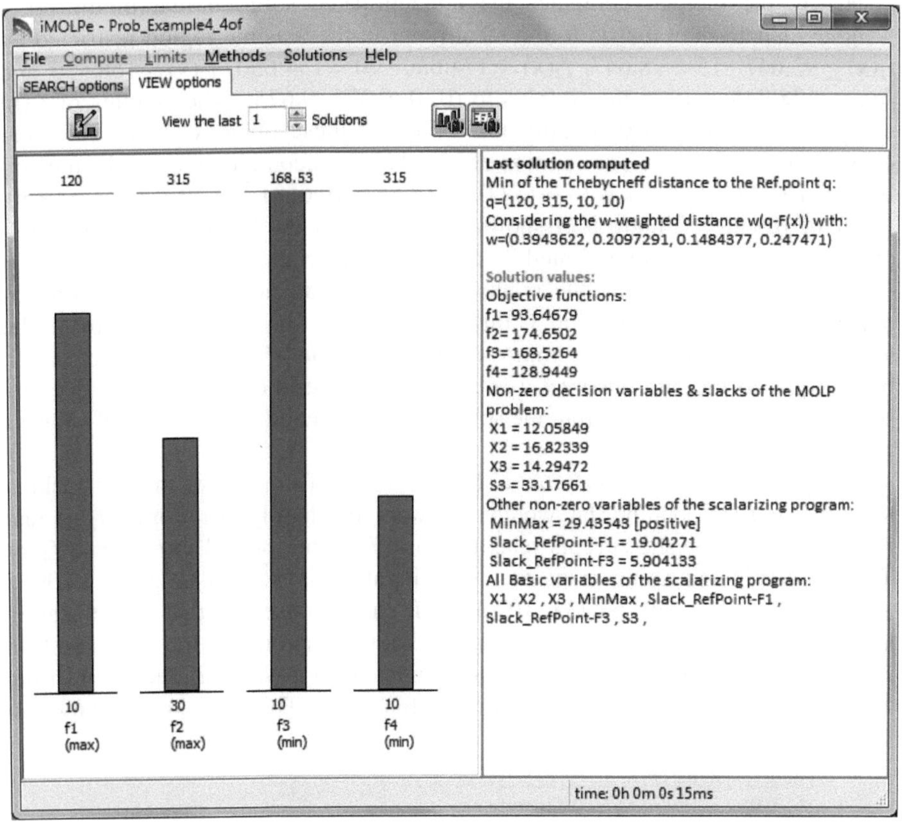


**Fig. 5.19** Example 4: nondominated solution obtained in the first iteration of the STEM method

information presented in Fig. 5.19 corresponds to the α in the description of the method in Chap. 4). In the first iteration the weights are (0.394, 0.210, 0.148, 0.247).

As previously mentioned (in Chap. 2) the worst value of the *pay-off* table for a given objective function generally does not correspond to the worst value of that objective function over the nondominated set (i.e., the nadir value). This situation is well illustrated in this problem, in which the worst value (maximum) of $f_3(\mathbf{x})$ in the *pay-off* table is 60 and $z_3 = 168.526$ in the last solution computed (Fig. 5.19).

The version of the STEM method implemented in iMOLPe is more flexible than the original one proposed by Benayoun et al. (1971) because it allows simultaneous relaxations of several objective functions and relaxing a given objective function more than once.

Suppose that the DM wants to perform a second iteration of the STEM method and he/she considers the values of $f_1(\mathbf{x})$ and $f_2(\mathbf{x})$ satisfactory in the solution obtained in the first iteration ($\mathbf{x}^5$, $\mathbf{z}^5$ in Fig. 5.19). Accordingly, he/she accepts relaxing 15 in $f_1(\mathbf{x})$ and 20 in $f_2(\mathbf{x})$ in order to try to improve $f_3(\mathbf{x})$ and $f_4(\mathbf{x})$ (i.e., to decrease their values)—Fig. 5.20.

Hence, the method computes the nondominated solution that minimizes a weighted Chebyshev metric to the ideal point in the feasible region restricted by $f_1(\mathbf{x}) \geq 93.647 - 15 = 78.647$, $f_2(\mathbf{x}) \geq 174.650 - 20 = 154.650$, $f_3(\mathbf{x}) \leq 168.526$ and $f_4(\mathbf{x}) \leq 128.945$, setting the weights to (0, 0, 0.375, 0.625). The weights of $f_1(\mathbf{x})$ and $f_2(\mathbf{x})$ are equal to 0 because the values of these objective functions were considered satisfactory. The solution obtained in the second iteration of the method is shown in Fig. 5.21; it is $\mathbf{x}^6 = (12.058, 18.642, 10.658)$, $\mathbf{z}^6 = (97.283, 154.650, 168.526, 105.309)$. Figure 5.21 shows bar graphs for the last two solutions computed, being the latter represented by the bars on the right.

It can be observed that the relaxation in $f_1(\mathbf{x})$ had no effect in the solution obtained in the second iteration as this objective function has even improved; the value of $f_4(\mathbf{x})$ decreased (i.e., improved) but the value of $f_3(\mathbf{x})$ did not get better, being equal to the one obtained in $\mathbf{z}^5$. It can be experimentally verified that any isolated relaxation of $f_1(\mathbf{x})$ from the solution of the second iteration ($\mathbf{x}^6$, $\mathbf{z}^6$) always results in the same solution.

Suppose that the DM performs a third iteration of the STEM method seeking to improve $f_3(\mathbf{x})$, but he/she would not want to sharply deteriorate $f_2(\mathbf{x})$ and $f_4(\mathbf{x})$. Therefore, he/she accepts relaxing 30 in $f_1(\mathbf{x})$ but only 2 in $f_2(\mathbf{x})$ and $f_4(\mathbf{x})$. The nondominated solution of the third iteration is $\mathbf{x}^7 = (0, 22.671, 12.091)$, $\mathbf{z}^7 = (102.774, 152.650, 57.433, 107.309)$. It can be observed that the relaxation allowed for $f_1(\mathbf{x})$ did not produce any effect, but the small relaxation of two units in $f_2(\mathbf{x})$ and $f_4(\mathbf{x})$ allowed $f_3(\mathbf{x})$ to have a significant improvement, decreasing from 168.526 to 57.433 (this is a minimizing function). This solution is shown in

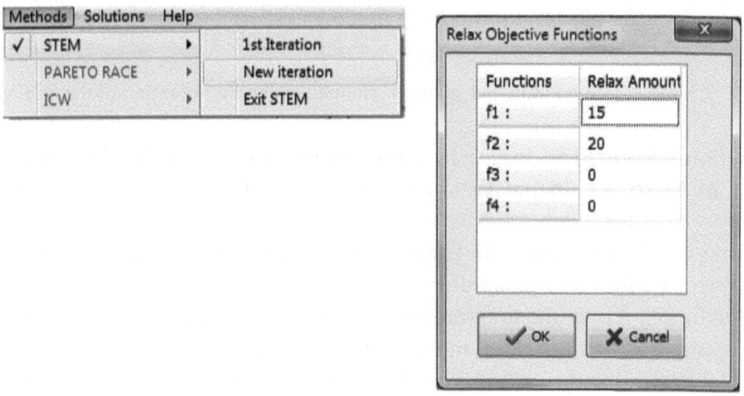


**Fig. 5.20** Example 4: specifying relaxation amounts for a new iteration of the STEM method

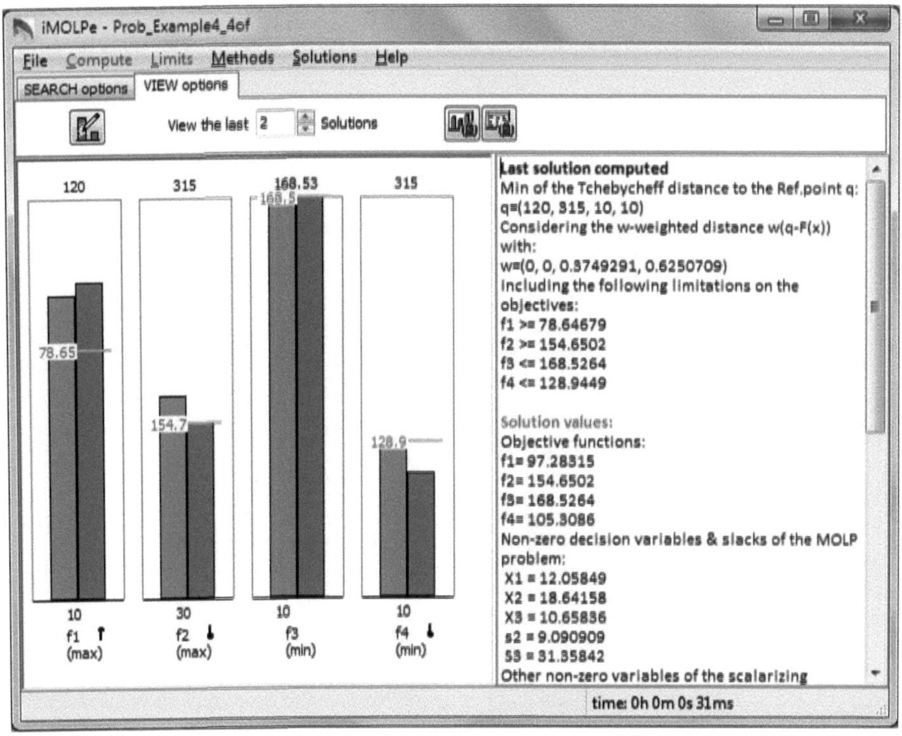


**Fig. 5.21** Example 4: nondominated solution obtained in the second iteration of the STEM method

Fig. 5.22, being represented by the darker bars on the right for each objective function.

If the DM decides to exit the STEM method to continue the search for new nondominated solutions using another procedure, he/she can keep or not the additional constraints imposed on the objective function values. If the DM accepts these constraints, he/she will always be allowed to discard them at any phase of the future search.

### 5.2.5 *Final Comments*

Examples 1, 2 and 3 aimed at illustrating some main features of the iMOLPe software using problems with two and three objective functions. As can be seen in these examples, the decomposition of the weight space together with the 2D/3D graph of the objective function space are very useful means for exploring the results, particularly for teaching and decision support purposes, as they allow to better understand the geometry of the nondominated frontier and the underlying trade-offs among the objective functions.

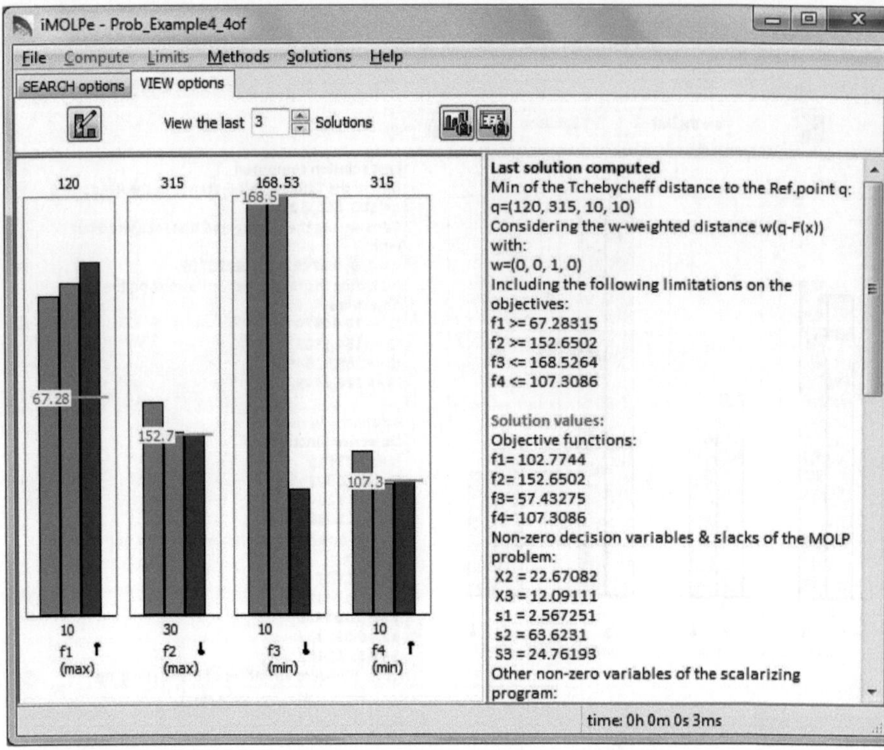


**Fig. 5.22** Example 4: nondominated solution obtained in the third iteration of the STEM method

The software can also deal with problems with four or more objective functions, as shown in Example 4, although the graphical representation of the objective space and the weight space are not available. Example 4 also aimed to illustrate the use of a more structured interactive procedure, the STEM method.

## 5.3 Proposed Exercises

1. Consider the following MOLP problem with three objective functions:

$$\text{Max} \begin{bmatrix} 3 & 1 & 2 & 1 \\ 1 & -1 & 2 & 4 \\ -1 & 5 & 1 & 2 \end{bmatrix} \begin{bmatrix} x_1 \\ x_2 \\ x_3 \\ x_4 \end{bmatrix} \quad \text{s.t.} \begin{bmatrix} 2 & 1 & 4 & 3 \\ 3 & 4 & 1 & 2 \\ 1 & 2 & 3 & 4 \end{bmatrix} \begin{bmatrix} x_1 \\ x_2 \\ x_3 \\ x_4 \end{bmatrix} \leq \begin{bmatrix} 60 \\ 60 \\ 50 \end{bmatrix}$$

$$\mathbf{x} \geq \mathbf{0}, \mathbf{x} \in \mathbb{R}^4$$

Use the iMOLPe Software to:

(a) Compute all nondominated basic solutions to the problem.
(b) Analyze the decomposition of the weight space to determine

- the nondominated edges and faces;
- the vertices that define each nondominated face;
- whether any of the objective functions reach its optimal value in more than one nondominated solution;
- whether there is any nondominated solution that provides the optimal value simultaneously for more than one objective function.

2. Consider the following MOLP problem with two objective functions:

$$\begin{array}{lll} \min & f_1(\mathbf{x}) = x_1 + 2x_2 + 3x_3 & \\ \max & f_2(\mathbf{x}) = -x_1 + 3x_2 + 8x_3 & \\ \text{s.t.} & x_1 + x_2 + x_3 & \leq 150 \\ & x_1 + 5x_2 + 3x_3 & \geq 56 \\ & -x_1 + x_3 & \leq 10 \\ & x_1, x_2, x_3 \geq 0 & \end{array}$$

Use the iMOLPe Software to:

(a) Compute the nondominated solution that optimizes the weighted-sum of the objective functions with the weight vector $(\lambda_1, \lambda_2) = (0.6, 0.4)$. Does this solution optimize any of the objective functions of the bi-objective problem?
(b) Find an approximate range of weights that lead to the solution computed in a) by visually inspecting the weight space.
(c) Compute the nondominated basic solution adjacent to the solution computed in a) that improves $f_1(\mathbf{x})$ and worsens $f_2(\mathbf{x})$ in relation to the previous solution.
(d) Compute the nondominated solution that optimizes $f_1(\mathbf{x})$ restricting $f_2(\mathbf{x})$ to values not lower than 100.
(e) Compute the nondominated solution corresponding to the projection of the reference point (60, 120) onto the nondominated set.

3. Consider the following MOLP problem with three objective functions:

$$\begin{array}{llll} \max & z_1 = f_1(\mathbf{x}) = & x_1 & \\ \max & z_2 = f_2(\mathbf{x}) = & x_2 & \\ \max & z_3 = f_3(\mathbf{x}) = & x_3 & \\ \text{s.t.} & & x_1 + x_2 + x_3 & \leq 5 \\ & & x_1 + 3x_2 + x_3 & \leq 9 \\ & & 3x_1 + 4x_2 & \leq 16 \\ & & 0.4x_1 + 0.2x_2 + x_3 & \leq 4.7 \\ & & x_1 + 0.3x_2 + 0.3x_3 & \leq 4.5 \\ & & 0.15x_1 + x_2 + 0.15x_3 & \leq 2.8 \\ & & x_1, x_2, x_3 \geq 0 & \end{array}$$

Use the iMOLPe software to:

(a) Compute the nondominated solutions that individually optimize each objective function.
(b) Compute new nondominated vertices (basic solutions) by optimizing weighted-sums of the objective functions (by direct selection of weights) until all the vertices defining a face are identified. How do you identify that face on the weight space?
(c) Suppose that the DM only wants new nondominated vertices satisfying the following constraint on the objective function values: $f_2(\mathbf{x}) \geq 1.5$. Identify the region of the weight space satisfying this constraint and compute the nondominated vertices that are reachable considering this constraint.

4. Consider the following MOLP problem with three objective functions:

$$
\begin{aligned}
\max z_1 &= f_1(\mathbf{x}) = 3x_1 + x_2 + 2x_3 + x_4 \\
\max z_2 &= f_2(\mathbf{x}) = x_1 - x_2 + 2x_3 + 4x_4 \\
\max z_3 &= f_3(\mathbf{x}) = -x_1 + 5x_2 + x_3 + 2x_4 \\
\text{s.t.}& \\
& 2x_1 + x_2 + 4x_3 + 3x_4 \leq 60 \quad (1) \\
& 3x_1 + 4x_2 + x_3 + 2x_4 \leq 60 \quad (2) \\
& x_1 + 2x_2 + 3x_3 + 4x_4 \leq 50 \quad (3) \\
& 4x_1 + 3x_2 + 2x_3 + x_4 \leq 80 \quad (4) \\
& 4x_1 + 5x_2 + 2x_3 + 3x_4 \leq 70 \quad (5) \\
& -x_1 + 4x_2 + 8x_3 + 5x_4 \leq 60 \quad (6) \\
& x_1, x_2, x_3, x_4 \geq 0
\end{aligned}
$$

Use the iMOLPe software to:

(a) Obtain the nondominated solutions that individually optimize each objective function and the corresponding indifference regions on the weight space.
(b) Comment on the following statement: "There is at least another nondominated basic solution that is alternative optimal for $f_k(\mathbf{x})$, $k = 1, 2, 3$".
(c) Compute another nondominated solution (and the respective indifference region) by optimizing a weighted-sum of the objective functions with a combination of weights such that $\lambda_1 \leq \lambda_2 \leq \lambda_3$.
(d) Compute another nondominated solution (and the respective indifference region), which could be obtained by a pivoting operation on the simplex tableau associated with the nondominated solution obtained in c).
(e) Characterize all nondominated vertices, edges and faces.
(f) Comment on the following statement: "There are nondominated vertices to the problem that are dominated in all the three bi-objective problems that are formed by dropping one of the objective functions (i.e., $f_1(\mathbf{x})$ and $f_2(\mathbf{x})$, $f_1(\mathbf{x})$ and $f_3(\mathbf{x})$, $f_2(\mathbf{x})$ and $f_3(\mathbf{x})$".

Add the following objective function to the problem:

$$\min z_4 = f_4(\mathbf{x}) \; = x_1 + x_2 + x_3 + x_4$$

(g) Obtain a nondominated solution using the STEM method.
(h) Suppose that the DM considers satisfactory the value $z_4$ in the previous solution and accepts relaxing this function by 2. What is the solution obtained in the second iteration of the STEM method? Which objective functions improved their values with respect to the previous solution?
(i) Exit the STEM method but keep the constraints imposed on the objective functions. Optimize a weighted-sum of the objective functions considering all weights equal and including those additional constraints. What is the solution obtained? Which functions improved and which worsened their values in relation to the solution computed in h)?

5. Consider the MOLP problem of Exercise 4 with three objective functions, but change the right-hand-side of constraint (4) from 80 to 50.

   (a) Compute all nondominated basic solutions to the problem.
   (b) Identify a degenerate solution and all regions on the weight space associated with the efficient bases of this solution.
   *Hint*: the number of efficient bases is 4 and there are overlapping regions on the weight space.

# Chapter 6
# Multiobjective Integer and Mixed-Integer Linear Programming

## 6.1 Introduction

The introduction of discrete variables into multiobjective programming problems leads to all-integer or mixed-integer problems that are more difficult to tackle, even if they have linear objective functions and constraints. The feasible set is no longer convex, and the additional difficulties go beyond those of changing from single objective linear programming to integer programming. Thus, in many cases the problems cannot be handled by adaptations of MOLP methods to deal with integer variables. In addition, there are approaches specifically designed for multiobjective pure integer problems that do not apply to the multiobjective mixed-integer case. Therefore, even for the linear case, techniques for dealing with multiobjective integer/mixed-integer linear programming (MOILP/MOMILP) involve more than the combination of MOLP approaches with integer programming techniques. In this chapter we focus on MOILP/MOMILP problems formulated as (6.1):

$$\left.\begin{array}{l} \max\ z_1 = f_1(\mathbf{x}) = \mathbf{c}_1\mathbf{x} \\ \qquad \cdots \\ \max\ z_p = f_p(\mathbf{x}) = \mathbf{c}_p\mathbf{x} \end{array}\right\} \quad \text{“Max”}\ \mathbf{z} = \mathbf{f}(\mathbf{x}) = \mathbf{C}\mathbf{x}$$
$$\text{s.t} \qquad \mathbf{x} \in X = \left\{\mathbf{x} \in \mathbb{R}^n : \mathbf{A}\mathbf{x} = \mathbf{b}, \mathbf{x} \geq \mathbf{0}, x_j \in \mathbb{N}_0, j \in I\right\} \tag{6.1}$$

where $I$ is the set of indices of the integer variables, $I \subseteq \{1,\ldots,n\}$, $I \neq \varnothing$. It is assumed that $X$ is bounded and non-empty. Let $Z$ denote the feasible region in the objective space, that is, $Z = \mathbf{f}(X)$. If all decision variables are integer then the multiobjective problem is all-integer (MOILP), which is a special case of the multiobjective mixed-integer case. In what follows we will refer to MOMILP as the general case, in which integrality constraints are imposed on all or a subset of the decision variables. For basic concepts concerning this type of problems, namely the characterization of efficient/nondominated solutions and the distinction



C. Henggeler Antunes et al., *Multiobjective Linear and Integer Programming*, EURO Advanced Tutorials on Operational Research,
DOI 10.1007/978-3-319-28746-1_6

between supported and unsupported nondominated solutions, please refer to Chap. 2.

As in other multiobjective programming problems, methods to address MOMILP problems may assume that the decision maker (DM)'s preferences are expressed a posteriori—*generating* methods—or be *interactive* methods. Generating methods are designed to find the whole set of nondominated solutions or a predefined subset, e.g., all supported or all extreme nondominated solutions. Interactive methods are characterized by phases of human intervention alternated with phases of computation.

Generating methods generally require a huge computational effort because all nondominated solutions, which may be a very large number even for moderate size problems, should be computed. For this reason, many generating methods are only intended for problems with binary variables (as it is easier to use enumeration techniques in these type of problems) or for bi-objective problems. A number of generating methods was firstly developed in the 1970s and 1980s decades, whereas the research focus shifted to interactive methods in the following decades. Still, new research on the field of exact generating methods has been carried out in recent years.

Several reviews of MOMILP methods have been published in the last three decades. Teghem and Kunsch (1986) presented a survey of interactive methods published until the final of 1985 (the first method dates from 1980). Covering the same period of time, another review is due to Rasmussen (1986) concerning multiobjective 0–1 programming, including both interactive and non-interactive methods. Clímaco et al. (1997) proposed a categorization of methods for multiobjective integer, linear and non-linear programming. Alves and Clímaco (2007) presented a review of interactive methods for MOMILP in which about twenty interactive methods are characterized and summarized.

## 6.2 Generating Methods and Scalarizing Processes

The first generating methods for MOILP/MOMILP problems were proposed in the 1970s and in the early years of the 1980s decade. Bitran (1977, 1979), Kiziltan and Yucaoglu (1983), and Deckro and Winkofsky (1983) proposed implicit enumeration algorithms for MOILP problems with binary variables only. The method of Bitran uses a constructive process, in which new nondominated solutions are successively generated and added to the set of nondominated solutions. This type of generating methods can be stopped before computing all nondominated solutions, returning a subset of the nondominated solution set. On the other hand, the methods of Kiziltan and Yucaoglu (1983) and Deckro and Winkofsky (1983) generate candidate solutions for nondominated solutions and the true nondominated set is known only at the end of the process. This type of methods operate, in the intermediate phases of the process, with *potentially* nondominated solutions, i.e., solutions that are not dominated by any solution already known. These methods cannot be interrupted before the end as they can fail in yielding nondominated solutions.

Klein and Hannan (1982) developed a constructive process for MOILP problems with general integer variables. It progressively restricts the feasible region through the introduction of auxiliary constraints, which eliminate nondominated solutions already calculated and solutions dominated by them. The method starts by optimizing one of the objective functions in the original feasible region. Then, it follows an iterative process in which the same objective functions is optimized in a feasible region restricted by additional constraints that force the next solution to be better in *some* objective function ('∨' conditions) with respect to *all* non-dominated points already known ('∧' conditions). Since the formulation of '∨' conditions as linear constraints needs auxiliary binary variables, the size of the auxiliary program increases from one iteration to the next. The process ends when the feasible region of the auxiliary program becomes empty. This method has been the basis for further developments of MOILP generating methods. Sylva and Crema (2004) presented a variation of the Klein and Hannan's algorithm considering the maximization of a weighted-sum of the objective functions instead of choosing only one of the objective functions to optimize in each iteration. In order to discuss these methods in more detail, let us first recall some MOMILP fundamental concepts concerning the optimization of weighted-sums or just one of the objective functions, considering additional constraints in all or some objective functions.

Consider the scalarizing program (6.2) that optimizes a *weighted-sum of the objective functions with additional constraints on the objective functions*, where $\boldsymbol{\lambda} \in \Lambda = \{\boldsymbol{\lambda} \in \mathbb{R}^p: \lambda_k > 0, k = 1,\ldots,p, \sum_{k=1}^{p} \lambda_k = 1\}$.

$$\begin{aligned} &\max \sum_{k=1}^{p} \lambda_k f_k(\mathbf{x}) \\ &\text{s.t. } f_k(\mathbf{x}) \geq e_k, \qquad k = 1,\ldots,p \\ &\quad\ \mathbf{x} \in X \end{aligned} \tag{6.2}$$

The introduction of additional constraints on the objective function values into the weighted-sum scalarizing program (6.2) enables to reach any nondominated solution, supported or unsupported. Remind that the optimization of a weighted-sum in the original feasile region only enables to compute supported nondominated solutions to a MOMILP (all-integer or mixed-integer) problem (cf. Sect. 3.2).

Problem (6.2) can be regarded as a particularization of the general scalarization proposed by Soland (1979):

$$\begin{aligned} &\max\ g(f_1(\mathbf{x}),\ldots,f_p(\mathbf{x})) \\ &\text{s.t. } f_k(\mathbf{x}) \geq e_k, \qquad k = 1,\ldots,p \\ &\quad\ \mathbf{x} \in X \end{aligned} \tag{6.3}$$

Let $g(.)$ be an arbitrary real-valued function defined on $\mathbb{R}^p$ which is strictly increasing on Z. Then, $\mathbf{x}'$, $\mathbf{z}' = \mathbf{f}(\mathbf{x}')$ is an efficient/nondominated solution to the multiobjective problem if and only if it solves the problem (6.3) for at least one **e** vector. This proposition is also valid for problem (6.2).

The *e-constraint* scalarization, in which one of the objective functions of the multiobjective problem is optimized while the other objective functions are considered as constraints (cf. Sect. 3.1), can also be encompassed by the previous scalarization (both (6.2) and (6.3)). The weight assigned to the objective function selected to be optimized, say $f_i(\mathbf{x})$, is 1 and the other weights are 0 or a very small positive value ρ in order to ensure that a nondominated solution is computed rather than just a weakly nondominated solution, i.e.,: max $\left(f_i(\mathbf{x}) + \rho\sum_{k=1,k\neq i}^{p} f_k(\mathbf{x})\right)$. The *e-constraint* scalarizing problem (6.4) enables to compute every nondominated solution to the MOMILP problem.

$$\begin{aligned} \max\ & f_i(\mathbf{x}) + \rho \sum_{k=1,k\neq i}^{p} f_k(\mathbf{x}) \\ \text{s.t. } & f_k(\mathbf{x}) \geq e_k, \qquad k = 1,\ldots,p, k \neq i \\ & \mathbf{x} \in X \end{aligned} \tag{6.4}$$

Although the previous scalarizing problems enable to fully characterize the nondominated set of a MOMILP problem, they do not offer direct means to compute every nondominated solution and to ensure that all of them have been computed. They are only scalarizing techniques for computing nondominated solutions, which have been used in several methods, either in the exposed forms or variants thereof. Other scalarizing techniques used in MOMILP will be discussed later in this chapter.

The methods of Klein and Hannan (1982) and Sylva and Crema (2004) referred to above are intended to compute all nondominated solutions to a MOILP problem. The former considers the objective function of the *e-constraint* scalarization (6.4), while the latter uses a *weighted-sum* objective function as in (6.2). However, they require more than $p$ additional constraints in order to ensure, in each iteration, that the next solution is different from the previous ones and all nondominated solutions have been computed at the end of the process. Accordingly, after finding each nondominated point, these methods require more $p$ new auxiliary binary variables and $p+1$ new constraints as an operational means to guarantee that a new nondominated point to the MOILP problem will be computed. The algorithm stops when the scalarizing problem becomes infeasible.

Let us briefly describe the general principle of these algorithms. Suppose that the method will perform the iteration $h+1$, so it has already computed $h$ different nondominated points $\mathbf{z}^t = (z_1^t, z_2^t, \ldots, z_p^t)$, $t = 1,\ldots,h$; the binary variables $y_{tk}$ and the following constraints are added to $\mathbf{x} \in X$ in order to find the $(h+1)^{th}$ nondominated solution:

$$f_k(\mathbf{x}) \geq (z_k^t + 1)y_{tk} - \mathrm{M}(1 - y_{tk}), \quad k = 1, \ldots, p; t = 1, \ldots, h$$

$$\sum_{k=1}^{p} y_{tk} \geq 1, \quad t = 1, \ldots, h$$

$$y_{tk} \in \{0, 1\}, \quad k = 1, \ldots, p; t = 1, \ldots, h$$

M is a sufficiently large positive constant so that the constraint *k,t* is redundant when $y_{tk} = 0$; the lower bound $(z_k^t + 1)$ is imposed to $f_k(\mathbf{x})$ when $y_{tk} = 1$. It is assumed that all objective functions are integer valued, so $f_k(\mathbf{x}) \geq (z_k^t + 1)$ imposes that the $k^{th}$ objective must be strictly greater than its value in $\mathbf{z}^t$ and no gap will remain between the value $z_k^t$ and the bound $z_k^t + 1$. The constraint $\sum_{k=1}^{p} y_{tk} \geq 1$ for each *t* forces the new nondominated solution to be better than the nondominated point $\mathbf{z}^t$ in at least one of the objectives; one of these constraints is imposed for each $\mathbf{z}^t$ previously computed.

Lokman and Köksalan (2013) proposed an improvement to this approach, decreasing the number of binary variables from $hp$ to $h(p-1)$ and additional constraints from $h(p+1)$ to $hp$ to find the $(h+1)^{th}$ new nondominated solution. However, the model size still grows and causes computational difficulties when the number of nondominated solutions is large. Theoretically, only $p-1$ additional constraints in the *e-constraint* scalarization (6.4) are sufficient to compute every nondominated solution, but the difficulty lies in determining appropriate lower bounds $e_k$. Based on this principle, Lokman and Köksalan (2013) proposed another algorithm to improve the previous one, which identifies the necessary constraints by solving submodels with $p-1$ or fewer additional lower bound constraints. Although this process avoids binary variables, it requires a significant number of submodels to be solved.

The difficulties in finding all nondominated solutions decrease substantially in bi-objective problems for which the design of an algorithm that scans the whole nondominated frontier is much easier, either for all-integer or mixed-integer multiobjective problems. An approach similar to the one of Klein and Hannan but restricted to MOILP problems with two integer valued objective functions was proposed by Chalmet et al. (1986). In each iteration, a weighted-sum of the objective functions is optimized with $p=2$ additional constraints on the objective function values–scalarization (6.2). No additional binary variables are required and the lower bounds $e_k$ can be easily determined for each new computation. The method works as follows.

Firstly, the nondominated solutions that optimize individually each objective function are computed and these solutions form a first pair to be analyzed. In each iteration, the method picks a pair of nondominated solutions already computed ($\mathbf{z}^a$, $\mathbf{z}^b$) that are candidate for being adjacent and analyzes whether there is any nondominated solution between them. It solves (6.2) with $f_k(\mathbf{x}) \geq \breve{z}_k + 1$, $k=1$, 2 where $\breve{z}_k = \min\{z_k^a, z_k^b\}$. Solutions $\mathbf{z}^a$ and $\mathbf{z}^b$ are really adjacent if the scalarizing problem (6.2) is infeasible—as illustrated in Fig. 6.1a; otherwise, a new nondominated solution between $\mathbf{z}^a$ and $\mathbf{z}^b$ is found, say $\mathbf{z}^c$, and the pairs $(\mathbf{z}^a, \mathbf{z}^c)$ and $(\mathbf{z}^c, \mathbf{z}^b)$ are formed to be further analyzed—this situation is illustrated in

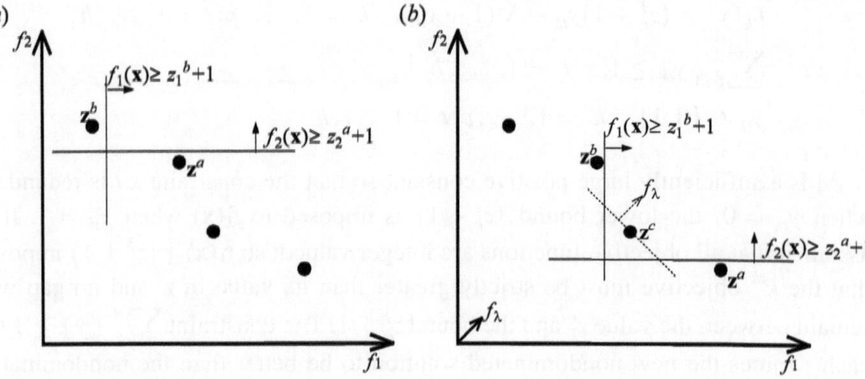


**Fig. 6.1** Computing all nondominated solutions in bi-objective MOILP problems using weighted-sums with bounds on the objectives

Fig. 6.1b; in this figure, $f_\lambda$ denotes the gradient of a weighted-sum of the objective functions with an arbitrary weight vector. The method stops when all pairs of solutions have been analyzed.

The *e-constraint* scalarization (6.4) can also be used to compute all nondominated solutions to a bi-objective integer problem. Once again, consider that the objective functions are integer valued, so that constraints of the type $f_k(\mathbf{x}) \geq \breve{z}_k + 1$ can be used to ensure that the $k^{th}$ objective function improves its value with respect to $\breve{\mathbf{z}}$ and no other objective values exist between $\breve{z}_k$ and $\breve{z}_k + 1$. If the objectives are real-valued, then $\breve{z}_k + 1$ can be replaced by $\breve{z}_k + \varepsilon$, with $\varepsilon$ a small positive value, and the method computes all nondominated solutions that differ from any other by (at least) $\varepsilon$ in each objective function.

The classical *e-constraint* generating method for bi-objective integer programming problems can be described as follows, where $Z_E$ denotes the set of nondominated solutions to the problem. Without loss of generality, let us assume that $i = 1$ is the index of the objective function chosen to be optimized in (6.4).

*Step 0:* $Z_E = \varnothing$
Set $e_2 = -M$

*Step 1:* Solve:

$$\begin{aligned} &\max\ f_1(\mathbf{x}) + \rho f_2(\mathbf{x}) \\ &\text{s.t.}\ \ f_2(\mathbf{x}) \geq e_2 \\ &\mathbf{x} \in X \end{aligned}$$

If the problem has no feasible solution, then STOP.
Otherwise, let the optimal solution be $\mathbf{x}'$ with $\mathbf{z}' = \mathbf{f}(\mathbf{x}')$.

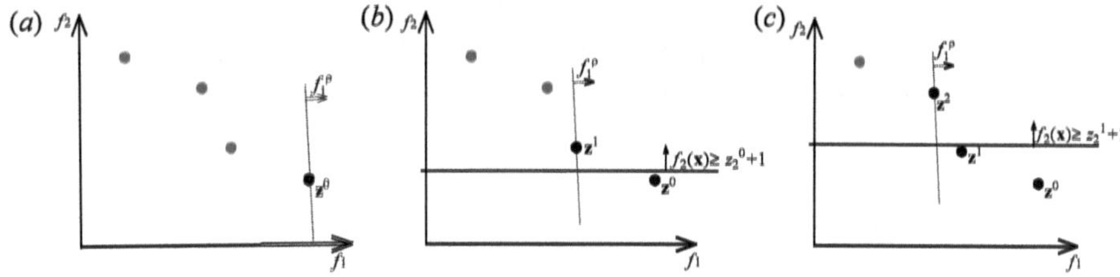

Fig. 6.2 Illustrating the *e-constraint* method for bi-objective integer problems

*Step 2:* $Z_E = Z_E \cup \{\mathbf{z}'\}$
Set $e_2 = z_2' + 1$
Return to *Step 1*.

M in *Step 0* is a large positive constant in order that the constraint $f_2(\mathbf{x}) \geq -M$ becomes redundant in the first optimization. As in (6.4), ρ is a small positive number (close to zero). Therefore, the first solution obtained in *Step 1* is a nondominated solution that optimizes $f_1(\mathbf{x})$; this solution presents the minimum value of $f_2(\mathbf{x})$. Nondominated solutions successively improving the second objective function are then computed. Figure 6.2 illustrates three iterations of the algorithm, respectively from (*a*) to (*c*), using the same example of Fig. (6.1).

So far we have been referring to methods that use scalarizing techniques to compute nondominated solutions based on the optimization of *weighted-sums with additional constraints on the objectives*, where the *e-constraint* scalarization may be included as a particular case. Other scalarizing techniques can be used, namely those based on *reference points*. Let us remind this type of scalarization (cf. Sect. 3.3) and further detail it in the particular context of MOILP and MOMILP problems.

Consider a point $\mathbf{z}^+$ of the objective space, called *reference point*, which satisfies $\mathbf{z}^+ \geq \mathbf{z}$ for all $\mathbf{z} \in Z$. The scalarizing problem (6.5) computes the nondominated solution that minimizes the distance to $\mathbf{z}^+$ according to a $\boldsymbol{\lambda}$-weighted Chebyshev metric, $\boldsymbol{\lambda} \geq \mathbf{0}$. A reference point often used is the *ideal point* (ideal solution) of the multiobjective problem, $\mathbf{z}^*$ (where $z_k^* = \max_{\mathbf{x} \in X} \{f_k(\mathbf{x})\}$, $k = 1, \ldots, p$) but other reference points may be used instead.

$$
\begin{aligned}
\min \quad & \left( \max_{k=1,\ldots,p} \left\{ \lambda_k \left( z_k^+ - f_k(\mathbf{x}) \right) \right\} \right) \\
\text{s.t.} \quad & \mathbf{x} \in X
\end{aligned}
\tag{6.5}
$$

If $\mathbf{x}' \in X$ is an efficient solution to a multiobjective mathematical program, then some $\boldsymbol{\lambda}' \geq \mathbf{0}$ exists such that $\mathbf{x}'$ optimizes (6.5) with $\boldsymbol{\lambda} = \boldsymbol{\lambda}'$ (Bowman 1976). Normalized weights are generally used, that is $\boldsymbol{\lambda} \in \Lambda_0 = \{\boldsymbol{\lambda} \in \mathbb{R}^p\colon \lambda_k \geq 0, k = 1, \ldots, p, \sum_{k=1}^{p} \lambda_k = 1\}$, and the previous assertion is still valid for $\boldsymbol{\lambda}' \in \Lambda_0$. Hence, the parameterization of this scalarizing problem on $\boldsymbol{\lambda} \in \Lambda_0$ enables to reach all efficient/nondominated solutions to the multiobjective problem, though it may also yield weakly nondominated solutions. This undesirable result is avoided by considering the *augmented weighted Chebyshev* programming problem formulated in (6.6), where ρ is a small positive constant.

$$\begin{aligned} \min \quad & \left( \max_{k=1,\ldots,p} \{\lambda_k (z_k^+ - f_k(\mathbf{x}))\} - \rho \sum\nolimits_{k=1}^{p} f_k(\mathbf{x}) \right) \\ \text{s.t.} \quad & \mathbf{x} \in X \end{aligned} \tag{6.6}$$

The term $\rho \sum\nolimits_{k=1}^{p} f_k(\mathbf{x})$ is a perturbation of the min-max objective function intended to ensure that the solution obtained is strictly nondominated. The *augmented weighted Chebyshev* programming problem can be written equivalently in the following away:

$$\begin{aligned} \min \quad & \left( v - \rho \sum\nolimits_{k=1}^{p} f_k(\mathbf{x}) \right) \\ \text{s.t.} \quad & \lambda_k (z_k^+ - f_k(\mathbf{x})) \le v, \; k = 1, \ldots, p \\ & \mathbf{x} \in X \\ & v \ge 0 \\ & \text{with } \boldsymbol{\lambda} \ge \mathbf{0} \text{ and } \mathbf{z}^+ \ge \mathbf{z} \text{ for all } \mathbf{z} \in Z \end{aligned} \tag{6.7}$$

Considering the problem (6.7) parameterized on $\boldsymbol{\lambda} \in \Lambda_0$ with a fixed reference point $\mathbf{z}^+ > \mathbf{z}^*$, there always exists a small enough ρ such that all nondominated solutions to a MOILP problem are reachable (Steuer and Choo 1983). In MOMILP problems (and also in nonlinear cases), even considering ρ very small there may be portions of the nondominated set (close to weakly nondominated solutions) that this scalarizing problem is unable to compute for a constant ρ. This case is illustrated in Fig. 6.3, where the segment from C to C′ is not reached. Nevertheless, ρ can be set so small that the DM is unable to discriminate those 'hidden' nondominated solutions from nearby weakly nondominated solutions (this corresponds to C′ getting closer to C in Fig. 6.3). It is worth noting that, even in the cases for which there is a ρ small enough that enables to reach any nondominated solution (the cases of MOLP and MOILP), the existence of such ρ is mainly of theoretical interest because it is not known a priori. Thus, we can say that, in practice, the scalarization (6.7) enables to characterize the whole nondominated set of a MOMILP problem.

Reference points $\bar{\mathbf{z}} \in \mathbb{R}^p$ that are attainable because they are inside the feasible region, or they are outside the feasible region but do not satisfy the condition $\bar{\mathbf{z}} \ge \mathbf{z}$ for all $\mathbf{z} \in Z$, can also be used providing that the $v$ variable in (6.7) is defined without sign restriction. This problem is defined in (6.8). It corresponds to the minimization of a distance from Z to the reference point if the latter is not attainable and to the maximization of such a distance otherwise. If the reference levels that compose the reference point are used as the controlling parameters, the (weighted) Chebyshev metric changes its form of dependence on controlling parameters and should be interpreted as an *achievement* scalarizing function (Lewandowski and Wierzbicki 1988).

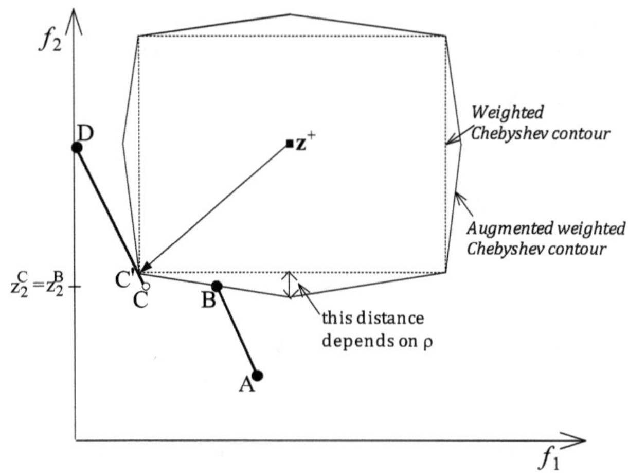


**Fig. 6.3** The augmented Chebyshev metric in MOMILP

$$
\begin{aligned}
\min\ & \left(v - \rho\sum\nolimits_{k=1}^{p} f_k(\mathbf{x})\right) \\
\text{s.t.}\ & \lambda_k(\overline{z}_k - f_k(\mathbf{x})) \le v,\ k = 1, \ldots, p \\
& \mathbf{x} \in X \\
& v \in \mathbb{R} \\
\text{with}\ & \boldsymbol{\lambda} \ge \mathbf{0} \text{ and } \overline{\mathbf{z}} \in \mathbb{R}^p
\end{aligned}
\tag{6.8}
$$

The optimal solution of the scalarizing problem (6.8) is a nondominated solution to the MOMILP problem for any $\overline{\mathbf{z}} \in \mathbb{R}^p$. The problem (6.8) may by parameterized on $\boldsymbol{\lambda}$, $\overline{\mathbf{z}}$, or both. If it is parameterized on $\overline{\mathbf{z}}$, the weights $\boldsymbol{\lambda}$ can be discarded (i.e., $\lambda_k = 1$ for all $k = 1,\ldots,p$) or fixed, thus playing the role of scale factors for normalizing purposes. There always exist reference points $\overline{\mathbf{z}} \in \mathbb{R}^p$ such that (6.8), with or without weights, produces a particular nondominated solution. This is still true considering only non-attainable reference points. Hence, there always exist reference points $\mathbf{z}^+ > \mathbf{z}^*$ such that (6.8) or (6.7) with $\overline{\mathbf{z}} = \mathbf{z}^+$ produces a particular nondominated solution. Figure 6.4a illustrates the outcome of the reference point scalarizing problem (6.7) or (6.8) when the weights are changed and Fig. 6.4b illustrates the outcome when the reference point is changed. The achievement scalarizing problem (6.8) should be considered in Fig. 6.4b rather than the augmented Chebyshev programming problem (6.7) as $v$ takes a negative value for the reference point $\overline{\mathbf{z}}^3$.

The principle of exploring new nondominated solutions between pairs of solutions for bi-objective MOILP problems, which has been described above for the method of Chalmet et al. (1986), has also been adopted by Solanki (1991) but using the augmented weighted Chebyshev metric (6.7). The method of Solanki applies to both all-integer and mixed-integer bi-objective linear problems and it is an extension for MOMILP of the Non-Inferior Set Estimation (NISE) method developed by

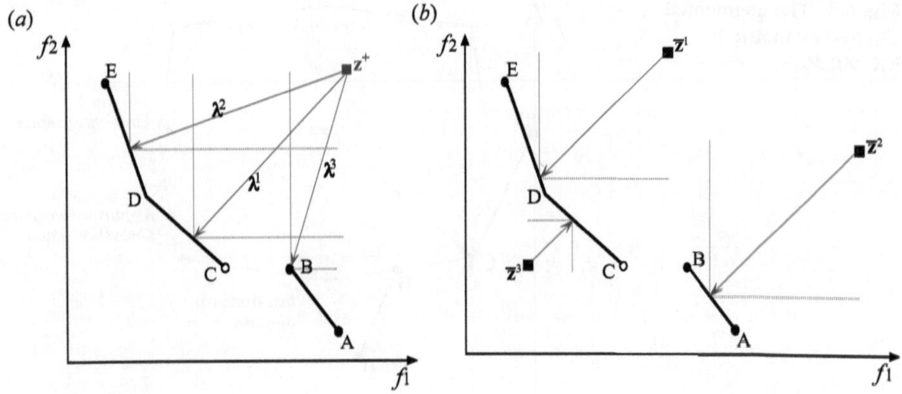


**Fig. 6.4** Varying the weights (**a**) or varying the reference point (**b**) in reference point scalarizing problems

Cohon et al. (1979) for bi-objective MOLP problems. The NISE method aims at obtaining a representation of the nondominated set by successively computing nondominated solutions that optimize weighted-sums of the objective functions. In NISE, the segment joining a pair of solutions, say $\mathbf{z}^a$ and $\mathbf{z}^b$, can be considered a good approximation of the nondominated frontier between $\mathbf{z}^a$ and $\mathbf{z}^b$ if the *error* of the approximation is within a predefined error bound. The measure of the *error* in MOLP is based on the convexity of the feasible region, which is no longer valid in integer or mixed-integer programming problems. Moreover, the weighted-sum problem used in NISE to generate nondominated solutions cannot obtain unsupported solutions to the MOMILP problem. These difficulties led Solanki to adopt the augmented weighted Chebyshev programming problem and to reformulate the measure of the *error*. In each iteration, the method changes both the reference point and the weights in (6.7), in order to explore intermediate solutions between nondominated solutions. This method works as follows.

Let $(\mathbf{z}^a, \mathbf{z}^b)$, with $z_1^a > z_1^b$, be a pair of nondominated points already computed. For this pair, an *error* is calculated as $\delta_{ab} = \max\left\{\left(z_1^a - z_1^b\right)/R_1, \left(z_2^b - z_2^a\right)/R_2\right\}$ where $R_1$ and $R_2$ are scale factors given by the difference between the maximum and the minimum of each objective function in the nondominated set. If $\delta_{ab} > \delta^{max}$ (the predefined error bound), then the scalarizing problem (6.7) is solved in order to search for a nondominated solution between $\mathbf{z}^a$ and $\mathbf{z}^b$. To explain how $\mathbf{z}^+$ and $\lambda$ are defined in each iteration, consider the example in Fig. 6.5. A rectangle formed with $\mathbf{z}^a$ and $\mathbf{z}^b$ as its corners is defined and $Diag(\mathbf{z}^a, \mathbf{z}^b)$ is the line passing through the appropriate corners of the rectangle. The reference point $\mathbf{z}^+$ is a point on $Diag(\mathbf{z}^a, \mathbf{z}^b)$ such that $\mathbf{z}^+ > \mathbf{z}^*$ and $(\lambda_1, \lambda_2)$ are calculated so that the diagonal of the Chebyshev metric coincides with $Diag(\mathbf{z}^a, \mathbf{z}^b)$. In the example of Fig. 6.5, the nondominated solution returned by (6.7) is $\mathbf{z}^c$ and the new pairs $(\mathbf{z}^a, \mathbf{z}^c)$ and $(\mathbf{z}^c, \mathbf{z}^b)$ are formed. The associated *errors* $\delta_{ac}$, $\delta_{cb}$ are then calculated, replacing $\delta_{ab}$. In the general case, if the nondominated solution returned by (6.7) is one of $\mathbf{z}^a$ or $\mathbf{z}^b$, then $\delta_{ab}$ is set to 0. The approximation of the whole nondominated frontier is thus

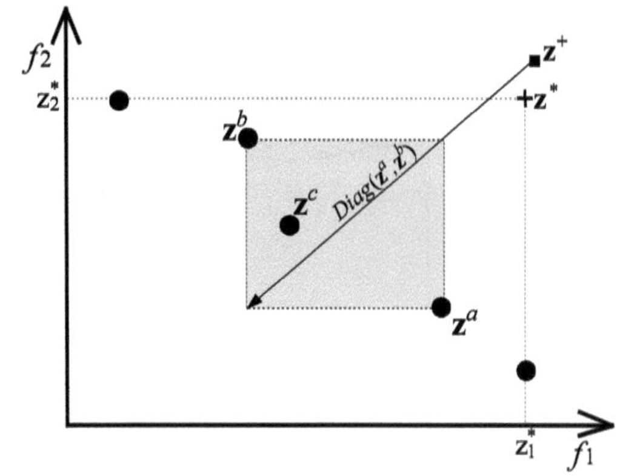


**Fig. 6.5** Illustration of the method of Solanki

progressively improved by decreasing the *errors* associated with the approximate representation of the pairs. In each iteration, the method chooses the pair of solutions with the largest *error* and stops when it is within the predefined *error* bound $\delta^{max}$. This method can be classified as a generating method. It can even compute all the nondominated solutions to a MOILP problem if $\delta^{max}$ is set to 0. In addition, the method can also be easily embodied in an interactive framework, in which the DM selects the pair of solutions to be analyzed in each iteration, and interactively decides whether the method is to be continued or stopped, without the need to define a priori the maximum error tolerance.

Still concerning generating methods, some other methods are worth of reference although we will not expose them in detail.

Mavrotas and Diakoulaki (1998, 2005) proposed a generating method for bi-objective mixed 0–1 linear problems. The technique consists in implicitly enumerating all possible values of the 0–1 variables, using a branch-and-bound algorithm, in order to generate potentially nondominated solutions. The dominated solutions are successively eliminated by pairwise comparisons and, at the final of the process, only nondominated solutions remain. Mavrotas and Diakoulaki consider only nondominated extreme points throughout the solution process and an improved method has been further proposed by Vincent et al. (2013).

Özlen and Azizoğlu (2009) proposed an extension for three objectives of the classical *e-constraint* method for bi-objective integer problems (described above). The approach intends to generate all nondominated solutions to the MOILP problem. An auxiliary bi-objective integer problem is defined: $\{\max f_1(\mathbf{x}) + \rho f_3(\mathbf{x}), \max f_2(\mathrm{x}) + \rho f_3(\mathbf{x})\colon \mathbf{x} \in X, f_3(\mathbf{x}) \geq e_3\}$. To generate the tri-objective nondominated set, the algorithm initially sets $e_3$ to a sufficiently small value (e.g., −M, as above) that does not cut any feasible solution, and generates all nondominated points to the surrogate bi-objective problem; then, it increases $e_3$ systematically and repeats the process in order to generate the other bi-objective nondominated sets. The $e_3$ value is updated as follows: $e_3 = z'_3 + 1$, with $z'_3 = \min\{f_3(\mathbf{x}), \mathbf{x} \in E\}$ where $E$ is the set of efficient solutions yielded by the previous surrogate bi-objective problem. An extension of this algorithm for more than three objective functions is also proposed but, as in other generating algorithms, its interest for $p > 3$ is mostly theoretical due to the

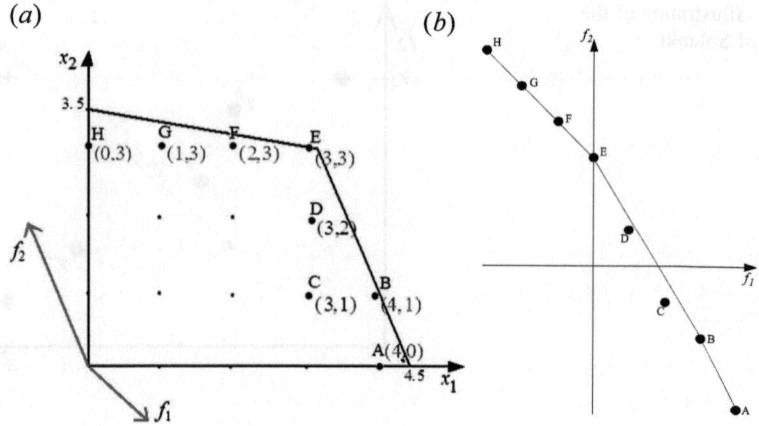


**Fig. 6.6** Illustrating extreme and non-extreme supported, and unsupported, nondominated solutions in MOILP

computational burden resulting from the recursive process. Also Kirlik and Sayın (2014) presented an algorithm for generating all nondominated solutions of MOILP problems based on the *e-constraint* scalarization.

In addition, there are some methods devoted to generate all *extreme supported* nondominated solutions to MOMILP problems (see also Chap. 2). A nondominated point $z' \in Z_E$ ($Z_E$ denotes the set of all nondominated points) is *supported* if it is located on the boundary of the convex hull of $Z$ (*conv Z*); *extreme supported* nondominated points $z' \in Z_E$ are vertices of *conv Z*. The optimization of a simple weighted-sum of the objective functions using the classical branch-and-bound method for (mixed-)integer linear programming yields an *extreme supported* nondominated solution. If there are alternative optima, a further exploration of the branch-and-bound tree allows computing non-extreme supported nondominated solutions. Figure 6.6 illustrates the difference between these types of solutions in a MOILP problem: (*a*) shows the problem in the decision space and (*b*) shows the corresponding nondominated solutions in the objective space, where the dotted line is just to evidence the boundary of the convex hull. Solutions A, B, E, F, G, H are supported nondominated solutions; among these, A, B, E, and H are the *extreme* solutions, and F and G are the *non-extreme* solutions. C and D are unsupported nondominated solutions. Concerning mixed-integer problems, observe, for instance, Fig. 6.3 above. In this problem, A, B and D are the *extreme supported* nondominated solutions. The other supported (but non-extreme) nondominated solutions are the points lying on the line segment [AB]; the points lying on ]CD[ are unsupported nondominated solutions and point C is a weakly nondominated solution.

Przybylski et al. (2010) and Özpeynirci and Köksalan (2010) proposed algorithms for determining all extreme supported nondominated points to MOMILP problems. These algorithms are mainly operationalized for three objective functions and they are based on the exploration of the weight space through the optimization of simple weighted-sums of the objective functions.

## 6.3 Interactive Methods

As already noticed, generating methods for problems with more than three objective functions are difficult to design and computationally expensive. In addition, if a large set of alternatives is presented to the DM at the final of the procedure, this will raise additional difficulties to the DM in analyzing all the information and making a final choice. Therefore, several researchers have developed *interactive* methods to deal with MOILP/MOMILP problems. Interactive methods enable to reduce the computational effort and aid the DM in the decision process. In interactive methods, the set of nondominated solutions is explored by a progressive articulation of the DM's preferences. This feature is shared by every interactive method, but there are different strategies followed by the authors. Some authors admit that the DM's preferences can be represented by an *implicit utility function* and the interactive process aims to 'discover' the optimum (or an approximation of it) to that implicit function. There are other approaches that allow a free progressive and selective search for nondominated solutions. These multiobjective approaches are not intended to converge to any 'best' solution, but to help the DM in the search for interesting nondominated solutions in order to identify a satisfactory compromise solution. We will refer to these approaches as *learning oriented* procedures.

There are also differences among the interactive methods in the type of scalarizing technique used to compute nondominated solutions, in the information provided to and required from the DM and in the way the method reduces the search space (not applicable to all methods). We present below a classification of several interactive methods representative of different approaches considering the following characteristics.

(a) Type of problems that the method is able to address: only multiobjective all-integer linear problems or general multiobjective mixed-integer linear problems.
(b) Scalarizing technique used to compute nondominated solutions: *weighted-sums*, *weighted-sums with additional constraints*, *e-constraint scalarization*, *reference point based scalarizing problems* or other technique.
(c) Information provided to and required from the DM, and the strategy used to reduce the scope of the search for new nondominated solutions.
(d) Type of protocol used to interact with the DM: the assumption of an *implicit utility function* or a *learning oriented* procedure.

In the following categorization we separate *bi-objective* methods from the *multiobjective* ones as the former group has naturally a more restricted application than multiobjective methods.

### 6.3.1 Bi-objective Interactive Methods

- Ramesh et al. (1990)
  - (a) Bi-objective all-integer linear programming.
  - (b) *Weighted-sums with additional constraints.*
  - (c) The method employs a modified version of the MOLP method of Zionts and Wallenius (1983) (see Chap. 4) using a branch-and-bound framework. The DM's preference structure is assessed using pairwise comparisons of nondominated solutions.
  - (d) Assumes an implicit *utility function* of the DM, pseudo-concave and non-decreasing.
- Aksoy (1990)
  - (a) Bi-objective mixed-integer programming.
  - (b) *e-constraint scalarization.*
  - (c) The method employs a branch-and-bound scheme to divide the subset of nondominated solutions of each node into two disjoint subsets by bisecting the range of values of a given objective function. The DM makes pairwise comparisons in order to determine the branching node and adjust the incumbent solution to the preferred nondominated solution.
  - (d) Assumes that the DM's preferences are consistent, transitive and invariant over the process aiming to optimize an implicit *utility function*.
- Shin and Allen (1994)
  - (a) Bi-objective all-integer (linear and nonlinear) problems, with concave objective functions and a convex feasible region (apart from the integrality constraints).
  - (b) A particular technique is used to compute, at each phase, the supported nondominated solution closer to an already known nondominated solution (to the right or to the left).
  - (c) The method successively excludes search regions by imposing constraints on the objective functions resulting from pairwise comparisons of nondominated solutions performed by the DM.
  - (d) Assumes an implicit *utility function* of the DM.
- Ferreira et al. (1994)
  - (a) Bi-objective mixed-integer programming.
  - (b) *Weighted-sums with additional constraints.*
  - (c) At each interaction the DM chooses a pair of solutions to further explore the nondominated region between them. The region to explore is reduced in relation to the original feasible region by constraints on the objective function values. The computation of new nondominated solutions enables to progressively eliminate some objective function regions, either by dominance or infeasibility.
  - (d) *Learning oriented* procedure.

### 6.3.2 Multiobjective Interactive Methods

- Villarreal et al. (1980), Karwan et al. (1985), Ramesh et al. (1986)

  (a) Multiobjective mixed-integer linear programming.
  (b) *Weighted-sums with additional constraints.*
  (c) These methods are extensions to MOMILP of the MOLP method of Zionts and Wallenius (1983), like the above-mentioned bi-objective method of Ramesh et al. (1990). These methods start by applying the Zionts-Wallenius algorithm to the linear relaxation of the MOMILP problem, following then a branch-and-bound phase to obtain a (mixed-)integer solution according to the DM's preferences. These preferences are assessed using pairwise evaluations of decision alternatives and tradeoff analysis.
  (d) Assumes an implicit *utility function* of the DM.

- Marcotte and Soland (1980, 1986), White (1985)

  (a) Problems with a convex or discrete feasible set. It can handle MOILP problems but is not applicable to the mixed-integer case.
  (b) *Weighted-sums with additional constraints.*
  (c) The algorithm separates the multiobjective problem into sub-problems, in a branch-and-bound scheme, by introducing a constraint on one objective function for each sub-problem (the feasible regions of the descendant problems are not necessarily disjoint). The nondominated solutions are computed by optimizing weighted-sums over the feasible subset corresponding to each sub-problem. The DM is asked to designate the incumbent solution among the nondominated solutions found thus far, and a node will be fathomed if its ideal point is not preferred to the incumbent solution. White (1985) proposed a Lagrangean technique to narrow the bounds provided by the ideal points for each sub-problem in order to eliminate more nodes of the tree that would be uninteresting to the DM.
  (d) Assumes that the DM's preferences are stable, but not requires the existence of an implicit *utility function*.

- Gonzalez et al. (1985)

  (a) Multiobjective integer linear programming.
  (b) *Simple weighted-sums* (at the first stage to compute supported nondominated solutions), *weighted-sums with a particular additional constraint* (at the second stage to compute unsupported nondominated solutions).

(c) The information required from the DM in each interaction consists of selecting the least preferred solution from a set of solutions in order to keep always $p$ candidate nondominated solutions. The hyperplane that passes through those $p$ objective points is used to define both the weights and the additional constraint of the scalarizing problem that is used to compute the next solution.
(d) Assumes an implicit *utility function* of the DM.

- Durso (1992)

(a) Multiobjective mixed-integer linear programming.
(b) *Augmented Chebyshev programming problem* (varying the reference point and including additional bounds on the objective functions).
(c) This method is a modification of the method of Marcotte and Soland (1986). Unlike the former, Durso's method is also suitable for mixed-integer programming. It employs a branch and bound scheme considering progressively smaller portions of the nondominated set by imposing lower bounds on the objective values. In each interaction the DM is asked to select the tree node for analysis and his/her most preferred solution among the $p + 1$ nondominated solutions computed for this node, which are the individual optima plus a 'central' solution obtained by the Chebyshev scalarizing problem. The selected solution is used to partition the problem into sub-problems, each one further restricting the bound of one objective. The reference point for the Chebyshev scalarizing problem is the ideal point of each node.
(d) *Learning oriented* procedure.

- L'Hoir and Teghem (1995)

(a) Multiobjective mixed-integer linear programming.
(b) *Augmented weighted Chebyshev programming problem* (varying the weights and the reference point, and including additional bounds on the objective functions).
(c) This method (called MOMIX) uses an interactive branch-and-bound framework related to the one introduced by Marcotte and Soland (1986). However, in the present case, the feasible regions of the descendant nodes of a given node are disjoint sets (contrariwise to the Marcotte and Soland's method). For each node, a nondominated solution is determined as in the STEM method (Benayoun et al. 1971) (see Chap. 4), i.e., by minimizing a weighted Chebyshev function to the ideal point of that node. In each interaction the DM is asked to choose the objective he/she wishes to improve with higher priority; this information is used to create the next sub-node.
(d) *Learning oriented* procedure.

- Vassilev and Narula (1993), Narula and Vassilev (1994)

(a) Multiobjective mixed-integer linear programming.

(b) *Achievement scalarizing problem* (varying the reference point and including additional constraints on the objective functions).
(c) In each interaction the DM is asked to specify a new reference point; its components should be better than the values in the last computed nondominated solution for the objectives the DM wishes to improve, worse for the objectives the DM accepts to deteriorate and equal for the others. This reference point is used to set the parameters for the next achievement scalarizing problem such that the next solution should move as far as possible from the previous solution in the objectives the DM wishes to improve with lower bounds (or equality constraints, depending on the preferences expressed by the DM) for the other objectives. The method does not consider any scheme to progressively reduce the search space.
(d) *Learning oriented* procedure.

- Karaivanova et al. (1995)

  (a) Multiobjective mixed-integer linear programming.
  (b) *Achievement scalarizing problem* (varying the reference point and including additional bounds on the objective functions).
  (c) The underlying principle is close to the method of Vassilev and Narula, although it is implemented in a distinct way. It devises a different achievement scalarizing problem, which minimizes the largest normalized difference to the aspiration levels in the objective functions the DM wants to improve (rather than maximizing the smallest normalized difference to the last solution, which is employed by the Vassilev and Narula's method). Lower bounds on the other objective functions are imposed. In each interaction, the DM indicates a reference point with aspiration levels for the objectives to be improved and reservation levels (lower bounds) for the others.
  (d) *Learning oriented* procedure.

- Alves and Clímaco (2000)

  (a) Multiobjective mixed-integer linear programming.
  (b) *Augmented Chebyshev programming problem* (varying the reference point).
  (c) In each interaction the DM may specify a new reference point or just indicate the objective function he/she wishes to improve with respect to the previous nondominated solution. The method is mainly devoted to perform *directional searches*, in which the DM has only to indicate the objective to improve in each interaction. Then, the method automatically adjusts the next reference point through a sensitivity analysis iterative procedure. Besides choosing the objective function to be improved at each moment, the DM has also the possibility of imposing bounds on the objective functions in order to have further control over the directional searches.

This method follows the same principle as a previous all-integer method presented in Alves and Clímaco (1999). The two methods differ in the techniques used to solve the scalarizing problems and, consequently, in the sensitivity analysis procedure. While the first method uses cutting planes the second one uses branch-and-bound, which is far more effective.

(d) *Learning oriented* procedure.

There are other general-purpose interactive methods which are applicable to multiobjective integer and mixed-integer programming, including nonlinear cases, e.g., Steuer and Choo (1983) and STEM method (see Chap. 4), provided that the (mixed) integer scalarizing problems are solved by an appropriate technique.

In the next section we will present in more detail the interactive method of Alves and Clímaco (2000). Although this is an interactive method, it can also be used as a *generating* method for bi-objective integer and mixed-integer linear problems. This operating mode will also be exemplified. A software implementing the method accompanies this book. The software incorporates other tools to deal with MOMILP problems, e.g., the possibility of computing nondominated solutions using distinct scalarizing techniques (including weighted-sums and weighted-sums with additional bounds on the objective functions). A brief overview of the software is also provided in this chapter.

## 6.4 An Interactive Reference Point Method Using Branch-And-Bound: Performing *Directional Searches* in MOMILP

Consider the achievement scalarizing problem (6.8) with equal fixed weights (i.e., $\lambda_k = 1$, $k = 1,\ldots,p$), parameterized on the reference point $\bar{\mathbf{z}}$. Without loss of generality, we assume that $\bar{\mathbf{z}} \geq \mathbf{z}^*$. Note that the outcome of this scalarizing problem for a given $\bar{\mathbf{z}}$ is the same as for the reference point $\mathbf{z}^+ = \bar{\mathbf{z}} + \delta\mathbf{e}$, where $\mathbf{e}$ a vector of 1s and $\delta$ a positive scalar. Therefore, any reference point can be shifted to be above all feasible objective points, which enables to restrict the $v$ variable to non-negative values. This corresponds to the minimization of the *augmented* (*non-weighted*) *Chebyshev* distance to the reference point $\mathbf{z}^+$, that is, problem (6.7) without weights (i.e., $\lambda_k = 1$ for all $k$).

The reference point method of Alves and Clímaco (2000) uses this scalarizing problem to compute nondominated solutions. In (6.9) the scalarizing problem is stated with the constraints being reorganized so that the terms with variables are on the left-hand side and the reference point, $\mathbf{z}^+$, is on the right-hand side.

$$\begin{aligned} & \min \left( v - \rho \sum_{k=1}^{p} f_k(\mathbf{x}) \right) \\ & \text{s.t.} \quad f_k(\mathbf{x}) + v \geq z_k^+ \quad k = 1, \ldots, p \\ & \qquad \mathbf{x} \in X, v \geq 0 \end{aligned} \tag{6.9}$$

Any nondominated solution to the MOMILP problem can be reached using (6.9) for some $\mathbf{z}^+ \geq \mathbf{z}^*$ provided that ρ is set small enough. In addition, we can obtain solutions that improve a specific objective function with respect to a previous solution by increasing the respective component of the reference point leaving the other components unchanged. This change in the reference point leads to a parametric right-hand side scalarizing problem. Alves and Clímaco (2000) developed a post-optimality technique with sensitivity analysis that identifies ranges of the reference points leading to the same nondominated solution and uses the branch-and-bound tree that solved the previous scalarizing problem as a starting point to compute the next nondominated solution. This enables to change automatically the reference point during a *directional search* and to save time in computation phases.

### 6.4.1 Interactive Algorithm

*Step 0* [optional]. Compute the pay-off table of the MOMILP problem.

*Step 1.* Ask the DM to specify a reference point, $\mathbf{z}^+ \in \mathbb{R}^p$. At the first interaction, the ideal point of the MOMILP problem is proposed by default, or the ideal point of the linear relaxation of the problem if *Step 0* has not been performed.

Solve the Chebyshev scalarizing problem (6.9) using branch-and-bound to obtain a nondominated solution; if necessary, $\mathbf{z}^+$ is firstly adjusted in order to satisfy $\mathbf{z}^+ \geq \mathbf{z}^*$ by adding a constant amount to all the components of $\mathbf{z}^+$.

*Step 2.* If the DM is satisfied and does not want to compute more nondominated solutions, STOP. Otherwise, if the DM wants to perform a global search and is willing to indicate explicitly a new reference point, return to *Step 1*.

Else, go to *Step 3*.

*Step 3.* Ask the DM to choose an objective function he/she wishes to improve with respect to the previous nondominated solution. Let $f_j(\mathbf{x})$ be the objective function specified by the DM.

A *directional search* is carried out by considering reference points of the form $\left(z_1^+, \ldots, z_j^+ + \theta_j, \ldots, z_p^+\right)$ in (6.9), with increasing values of $\theta_j > 0$, in order to produce a sequence of nondominated solutions that successively improve $f_j(\mathbf{x})$. The computation of new solutions in this direction stops when the DM wishes or a nondominated solution that optimizes $f_j(\mathbf{x})$ is reached.

Return to *Step 2*.

The core of the algorithm is the *Step 3* and the way a *directional search* is performed. It consists of optimizing successive scalarizing problems (6.9) that only differ from each other in the right-hand side of the $j^{\text{th}}$ constraint (corresponding to the $j^{th}$ objective function). This task is performed by an iterative process with two main phases: ($S$) *sensitivity analysis* and ($U$) *updating the branch-and-bound tree*.

The sensitivity analysis ($S$) returns a parameter value $\theta_j^{\max} > 0$ such that the structure of the previous branch-and-bound tree remains unchanged for variations in $z_j^+$ up to $z_j^+ + \theta_j^{\max}$. This means that reference points $\left(z_1^+, \ldots, z_j^+ + \theta_j, \ldots, z_p^+\right)$ with $\theta_j \leq \theta_j^{\max}$ either lead to the same nondominated solution or to different nondominated solutions that are easily computed because they result from the same node of the branch-and-bound tree. In the latter case, distinct nondominated solutions may be computed for different parameter values below $\theta_j^{\max}$ and these solutions are presented to the DM, who can interactively control the proximity of solutions he/she wants to visualize. This situation arises when a continuous region of solutions with the same integer part is being explored in a mixed-integer problem. The branch-and-bound tree is then updated ($U$) for $\theta_j$ slightly above $\theta_j^{\max}$ and a nondominated solution is produced. It may happen that this solution is the same as the last one because the $\theta_j^{\max}$ returned by the sensitivity analysis can be only a lower bound for the true maximum value of the parameter. In that case (which occurs more often in all-integer than in mixed-integer models) the process automatically returns to ($S$). The iterative process ends when a new nondominated solution is obtained in ($U$), which is then presented to the DM.

Let us now give further details about this process.

Assume that the problem (6.9) was solved with $\mathbf{z}^+ = \left(z_1^+, \ldots, z_j^+, \ldots, z_p^+\right)$ yielding the nondominated solution $\mathbf{x}^{\text{o}}$, $\mathbf{z}^{\text{o}}$. Then, the DM chooses the objective function $f_j(\mathbf{x})$ to be improved in relation to $\mathbf{z}^{\text{o}}$ . The next nondominated solutions will be obtained by solving the scalarizing problem (6.9) parameterized on $\theta_j \geq 0$ (6.10), where the $j^{\text{th}}$ component of the reference point is increased. We will refer to this problem as $\text{P}(\theta_j)$.

$$\begin{aligned} f(\theta_j) &= \min\left(v - \rho\sum\nolimits_{k=1}^{p} f_k(\mathbf{x})\right) \\ \text{s.t.}\quad & f_k(\mathbf{x}) + v - s_k = z_k^+ \qquad k = 1, \ldots, p\,,\ k \neq j \\ & f_j(\mathbf{x}) + v - s_j = z_j^+ + \theta_j \\ & \mathbf{x} \in X \\ & v \geq 0, s_k \geq 0, k = 1, \ldots, p \end{aligned} \tag{6.10}$$

The $s_k$, $k=1,\ldots,p$ are the surplus variables of the $k$ constraints associated with the objective functions of the MOMILP problem.

Consider the branch-and-bound tree that solved the mixed-integer linear programming problem (6.10) for $\theta_j=0$, that is P(0).

Each node Q of the optimal branch-and-bound tree that solved P(0) corresponds to a linear sub-problem of P(0). Regarding the parametric scalarizing problem P($\theta_j$), the linear sub-problem associated with a node $Q^r$ can be formulated as (6.11), which we will refer to as $LP^r(\theta_j)$:

$$\begin{aligned} f^r(\theta_j) &= \min\left(v - \rho\sum\nolimits_{k=1}^{p} f_k(\mathbf{x})\right) \\ \text{s.t.}\quad & f_k(\mathbf{x}) + v - s_k = z_k^+ \qquad k = 1, \ldots, p,\ k \neq j \\ & f_j(\mathbf{x}) + v - s_j = z_j^+ + \theta_j \\ & \mathbf{x} \in X^r = \left\{\mathbf{x} \in \mathbb{R}^n : \mathbf{A}\mathbf{x} = \mathbf{b}, \mathbf{x} \geq \mathbf{0}, L_i^r \leq x_i \leq U_i^r, i \in I\right\} \\ & v \geq 0, s_k \geq 0, k = 1, \ldots, p \end{aligned} \tag{6.11}$$

In (6.11) some $L_i^r$ may be zero and some $U_i^r$ infinite.

In what follows let *integer solution* denote a solution to a sub-problem (6.11) that has integer values for all the integer-restricted variables in the multiobjective problem.

Let us analyze the effects of increasing $\theta_j$ on the leaf nodes of the current branch-and-bound tree.

***Infeasible sub-problems***

When a sub-problem $LP^r(\theta_j)$ is infeasible for $\theta_j=0$, it will remain infeasible for all $\theta_j>0$. This proposition holds because, for $\theta_j^2 > \theta_j^1$, the feasible set of $LP^r(\theta_j^1)$ contains the feasible set of $LP^r(\theta_j^2)$. Hence, all the infeasible nodes of the current tree do have not to be further considered.

***The behavior of each node of the tree***

The optimal objective value of a minimizing parametric right-hand-side linear programming problem is a piecewise linear convex function of the parameter. This is a well-known result for linear programming with particularities for

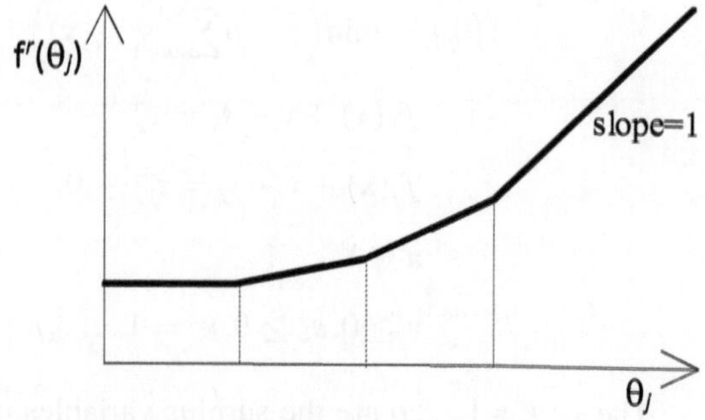


**Fig. 6.7** Example of the behavior of the optimal objective value of $LP^r(\theta_j)$

$LP^r(\theta_j)$: besides being convex, the piecewise linear function is non-decreasing and the slope of the function in the last interval of $\theta_j$ is 1 (see Fig. 6.7).

Let $\pi_k \geq 0$, $k = 1,\ldots,p$ be the dual variables associated with the first $p$ constraints of $LP^r(\theta_j)$. The function $f^r(\theta_j)$ must be non-decreasing because it is convex and $0 \leq \pi_j \leq 1$ (non-negative slopes). As $\theta_j$ grows through positive values, $LP^r(\theta_j)$ returns solutions with greater values for $f_j(\mathbf{x})$ until the solution that optimizes $f_j(\mathbf{x})$ in $X^r$ is reached. If $\theta_j$ increases more, $LP^r(\theta_j)$ will yield the same solution, say $\overline{\mathbf{x}}^r$, only varying the values of the variables $s_k$ and $v$. There exists a specific value of $\theta_j$ above which the Chebyshev distance ($v$) between the reference point and the image of $\overline{\mathbf{x}}^r$ in the objective space is exclusively given by the $j^{\text{th}}$ component, i.e., $s_j = 0$ and $s_k > 0$, $k \in \{1,\ldots,p\} \backslash \{j\}$. Hence, for $\theta_j$ larger than that specific value, $\pi_k = 0$, $k \in \{1,\ldots,p\} \backslash \{j\}$ and $\pi_j = 1$.

***(S) Sensitivity analysis to compute*** $\boldsymbol{\theta_j^{max}}$

Consider that the current nondominated solution $(\mathbf{x}^o, \mathbf{z}^o)$, which optimizes P(0), was produced by the node $Q^o$ (sub-problem $LP^o(0)$) of the branch-and-bound tree. The purpose of the *sensitivity analysis* is to provide a range of values $[0, \theta_j^{max}]$ for the parameter $\theta_j$ such that the optimal solutions of $P(\theta_j)$ will still be given by the node $Q^o$ and its current basis. We should note that the $\theta_j^{max}$ returned by this procedure may be lower than the true maximum value.

Two different situations may occur depending on whether $s_j$ is a basic variable in $LP^o(0)$ or not.

**(S.1)** $s_j$ is basic in $LP^o(0)$

If $s_j$ is basic in $LP^o(0)$, then for $\theta_j$ up to the current value of $s_j$, say $s_j^o$, neither the value of $f^o(\theta_j)$ changes (due to $\pi_j^o = 0$) nor $\mathbf{x}^o$; the node $Q^o$ is still the optimal node to $P(\theta_j)$ with $\theta_j \leq s_j^o$ and it yields the same nondominated solution $(\mathbf{x}^o, \mathbf{z}^o)$. Therefore, $\theta_j^{max} = s_j^o$ and there is no need to explore variations of $\theta_j$ under this value.

**(S.2)** $s_j$ is nonbasic in $LP^o(0)$

Let $\theta_j \in [0, \theta_j^{o^{max}}]$ be the positive interval of $\theta_j$ that ensures the optimality of the current basis for $LP^o(\theta_j)$. We will analyze only the case when variations of $\theta_j$

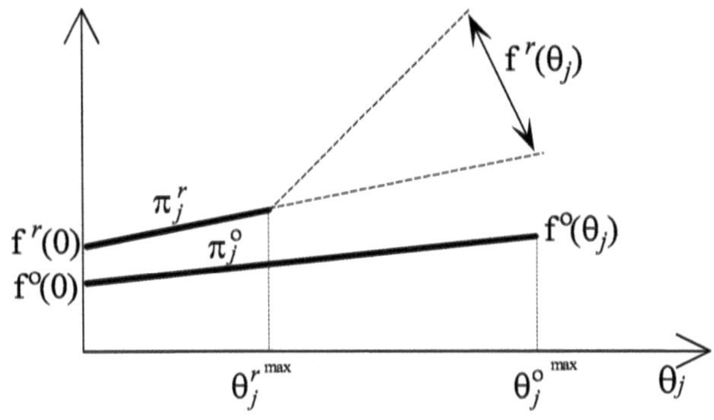


**Fig. 6.8** Example of a situation where $\pi_j^r \geq \pi_j^o$

within the same basis still produce integer solutions (otherwise, $\theta_j^{max} = 0$ is considered).

Under this condition, the node $Q^o$ provides feasible solutions to $P(\theta_j)$ for $\theta_j \leq \theta_j^{o\,max}$ and $f^o(\theta_j) = f^o(0) + \pi_j^o\, \theta_j$. The performance of node $Q^o$ when $\theta_j$ is increased will be compared with potential candidate terminal nodes of the branch-and-bound tree (containing or not an integer solution). Denoting by $\pi_j^r$ the current value of the dual variable associated with constraint $j$ in the node $Q^r$, the potential candidate nodes are the terminal nodes $Q^r$ that satisfy $\pi_j^r < \pi_j^o$. Nodes for which $\pi_j^r \geq \pi_j^o$ do not need to be considered because, for $\theta_j$ within the range $0 \leq \theta_j \leq \theta_j^{o\,max}$, they cannot provide solutions to $P(\theta_j)$ better than the one given by $LP^o$ $(\theta_j)$ (see Fig. 6.8). In fact, each linear segment of $f^r(\theta_j)$ has a slope between $\pi_j^r$ (bottom dashed line in Fig. 6.8) and 1 (top dashed line in Fig. 6.8). Since $\pi_j^r \geq \pi_j^o$ and $f^r(0) \geq f^o(0)$, then $f^r(\theta_j)$ cannot be lower than $f^o(\theta_j)$ for $0 \leq \theta_j \leq \theta_j^{o\,max}$.

For each potential node $Q^r$, an *intersection* parameter value $\theta_j^{o,r}$ is computed. Whereas $\theta_j^{o,r}$ is easily computed because it only requires information on current bases, it may be only a lower bound of the real $\theta_j$ for which $f^r(\theta_j)$ would intersect $f^o(\theta_j)$; $\theta_j^{o,r}$ represents a point where $f^r(\theta_j)$ *intersects* $f^o(\theta_j)$ given by a *'real' intersection* (e.g., Fig. 6.9a) or a *'virtual' intersection* (e.g., Fig. 6.9b, c and d):

$$\theta_j^{o,r} = \frac{f^r(0) - f^o(0)}{\pi_j^o - \pi_j^r}$$

For the sake of simplicity, thereafter the term *intersection* is also used with respect to nodes (e.g., "$Q^r$ *intersects* $Q^o$") for which $f^r(\theta_j)$ *intersects* $f^o(\theta_j)$ according to the description above.

The node $Q^o$ outperforms $Q^r$ at least until $\theta_j^{o,r}$ if $\theta_j^{o,r} < \theta_j^{o\,max}$—see examples in Fig. 6.9a and b—or at least until $\theta_j^{o\,max}$ if $\theta_j^{o,r} \geq \theta_j^{o\,max}$—see examples in Fig. 6.9c and d.

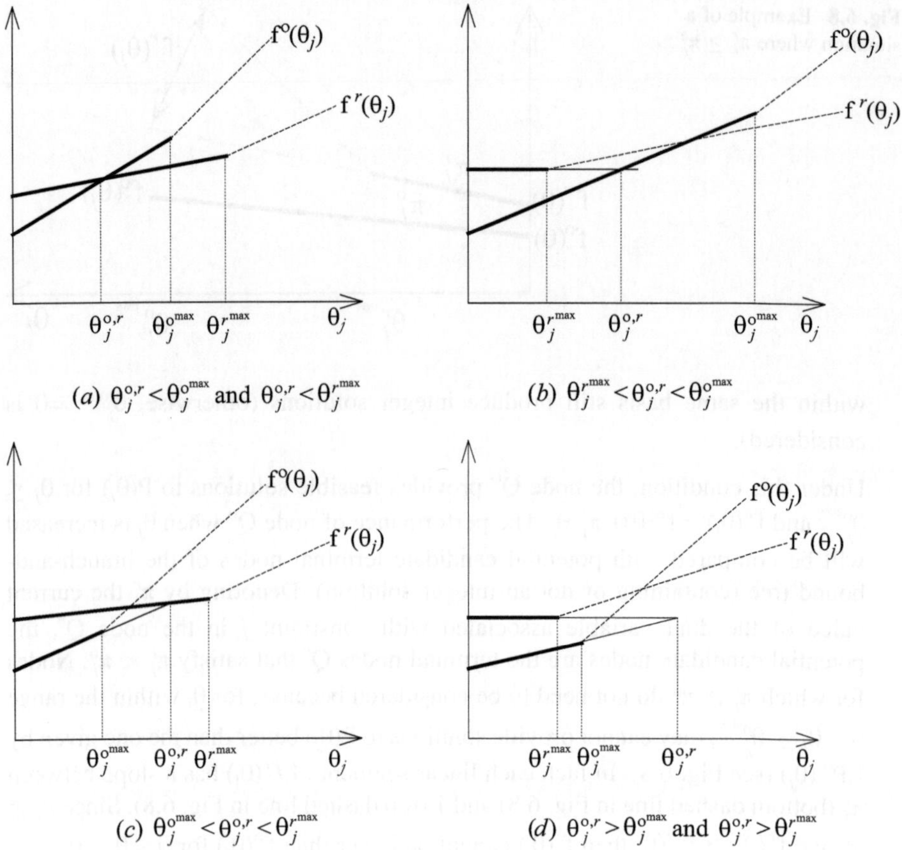


**Fig. 6.9** Examples of intersections. (**a**) $\theta_j^{o,r} < \theta_j^{o\max}$ and $\theta_j^{o,r} < \theta_j^{r\max}$, (**b**) $\theta_j^{r\max} < \theta_j^{o,r} < \theta_j^{o\max}$, (**c**) $\theta_j^{o\max} < \theta_j^{o,r} < \theta_j^{r\max}$, (**d**) $\theta_j^{o,r} > \theta_j^{o\max}$ and $\theta_j^{o,r} > \theta_j^{r\max}$

Hence, the parameter value that ensures the next nondominated solutions will still be given by node $Q^o$ is $\theta_j^{\max} = \min\left\{\theta_j^{o\max}, \min_r\left\{\theta_j^{o,r}\right\}\right\}$. The nondominated solutions that optimize $P(\theta_j)$ for $0 \le \theta_j \le \theta_j^{\max}$ can be obtained in a straightforward way by applying classic linear programming parametric analysis from the simplex tableau of $LP^o(0)$.

In situation (S.1), where the surplus variable $s_j$ is basic in $LP^o(0)$ and $\theta_j^{\max} = s_j^o$, no distinct nondominated solutions to the MOMILP problem can be found for $\theta_j \in [0, \theta_j^{\max}]$.

***Inactive nodes***

Any node $Q^r \neq Q^o$ for which $\pi_j^r = 1$ may be considered *inactive* while the parametric analysis refers to the $j^{\text{th}}$ constraint of the achievement scalarizing problem because it is not a potential node for *intersection*. As a result, $Q^r$ and its future descendants cannot provide the optimal solution to the scalarizing problem for larger values of $\theta_j$. The node will be activated if the DM changes the direction of search by choosing another objective function to be improved.

***Individual optima***

If $\pi_j^{o} = 1$ and all the other terminal nodes of the branch-and-bound tree are either *inactive* or their problems are infeasible, then the current nondominated solution ($\mathbf{x}^{o}$, $\mathbf{z}^{o}$) optimizes the objective function $f_j(\mathbf{x})$ of the MOMILP problem; if the optimum of $f_j(\mathbf{x})$ is reached, the directional search stops there.

***(U) Updating the branch-and-bound tree***

The previous branch-and-bound tree is used as a starting structure to solve the next scalarizing problems. For $0 \leq \theta_j \leq \theta_j^{max}$, the optimal solutions of $P(\theta_j)$ can be obtained in a straightforward way because they still optimize the parametric linear sub-problem associated with the node $Q^{o}$ of the tree. The structure of the tree does not change for this range of parameter values.

To continue searching for nondominated solutions in the same direction, the parameter is set to $\hat{\theta}_j = \theta_j^{max} + \varepsilon$, with $\varepsilon$ small positive. In MOILP problems with integer valued objective functions, $\hat{\theta}_j$ is set to the smallest integer value larger than $\theta_j^{max}$ because reference points can be restricted to integer components without loss of intermediate nondominated solutions.

The next reference point is thus $\left(z_1^+, \ldots, z_j^+ + \hat{\theta}_j, \ldots, z_k^+\right)$.

The procedure for updating the branch-and-bound tree begins by updating the simplex tableau of the node $Q^{o}$ for the new reference point. Afterwards, three different cases ought to be distinguished.

(U.1) The solution of the node $Q^{o}$ is no longer integer although its basis has not changed (this situation occurs within S.2 with $\theta_j^{max} = 0$).

After updating the information on the other terminal nodes, the branching process starts by splitting node $Q^{o}$ and the branch-and-bound proceeds as usual until the optimum of the scalarizing problem $P(\hat{\theta}_j)$ is reached.

(U.2) The node $Q^{o}$ has been *intersected* by another node and $\theta_j^{max}$ was given by the *intersection* parameter value $\theta_j^{o,r}$ for some node $Q^{r}$ in the sensitivity analysis phase (situation (S.2) with $\theta_j^{max} = \theta_j^{o,r}$—note that if there are several nodes that *intersect* node $Q^{o}$ within the range $[\theta_j^{max}, \hat{\theta}_j]$, all of them must be taken into account).

After updating the information on the intersecting node $Q^{r}$ for $\hat{\theta}_j$, $Q^{r}$ is compared again with $Q^{o}$: if $f^{o} \leq f^{r}$ (situation of *'virtual' intersection* like in Fig. 6.9b) then $Q^{o}$ still provides the optimal solution to the scalarizing problem; if $f^{o} > f^{r}$ then the solution of node $Q^{r}$ becomes the new optimal solution if it is integer; otherwise, $Q^{r}$ is the best candidate node and should be branched.

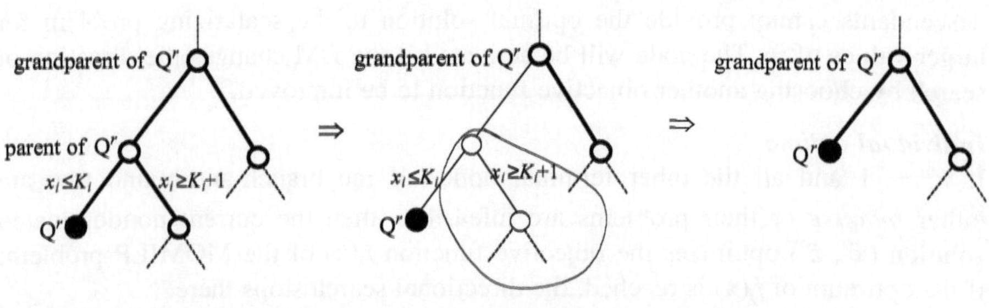


**Fig. 6.10** Low-level simplification process and the resulting tree

If Q″ needs to be branched, a *low-level simplification* of the tree is firstly attempted. The procedure begins by examining the branching constraint that links Q″ to its parent. If this constraint is no longer active (i.e., it became redundant as result of the basis change) then a simplification is made. This is called a low-level simplification because it only regards the lowest link from the node Q″ to its direct ancestor. Since no historical information is kept, this is the only branching constraint that was surely active in Q″ when the node was created. The simplification consists in the following: suppose that $x_i \le K_i$ is the inactive branching constraint that links Q″ to its parent (Fig. 6.10); the parent node is removed, as well as its descendants from the other branch ($x_i \ge K_i+1$); thus, Q″ becomes a direct descendant of its previous grandparent (Fig. 6.10). Note that Q″ will possibly be branched on the variable $x_i$. Without this pruning the procedure would lead to two consecutive branching constraints on the same variable.

Once the simplification is made, the information on the remaining terminal nodes is updated and the branch-and-bound proceeds as usual until the optimum of the scalarizing problem $P(\hat{\theta}_j)$ is reached.

(U.3) The basis of the node Q° changed, because $\theta_j^{max}$ was given by $\theta_j^{o^{max}}$ in the sensitivity analysis phase (situation S.1 and S.2 with $\theta_j^{max} = \theta_j^{o^{max}}$).

If the updated solution of Q° is integer, then the new nondominated solution is found. Otherwise, further branching is required, namely from Q° (the best candidate node). A *simplification* of the tree is first attempted, which now refers to *all* branching constraints that were active in the previous basis of node Q° and become redundant when the basis changed. Therefore, *low* (as in situation U.2) and/or *high-level simplifications* may occur, which are explained below. Then, the information on the remaining terminal nodes is updated and it is built for the new terminal nodes created by the simplification process. The branch-and-bound proceeds as usual until the optimum of the scalarizing problem $P(\hat{\theta}_j)$ is reached.

It should be stressed that if both situations (U.2) and (U.3) occur within the range $[\theta_j^{max}, \hat{\theta}_j]$, Q° must be compared again with the updated terminal nodes.

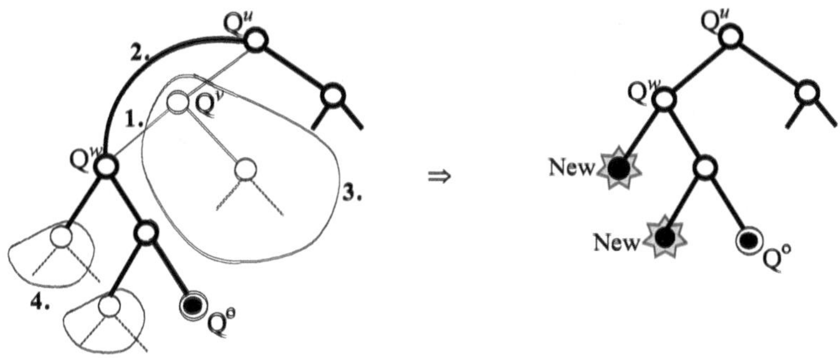


**Fig. 6.11** Illustration of the general simplification process

A *simplification* of the tree aims to avoid an evergrowing tree. The steps that rule the general simplification process in U.3 are the following (illustrated in Fig. 6.11). Recall that the simplification refers to all branching constraints that become redundant for $Q^o$ when its basis changed and $Q^o$ is the current best candidate node with a non-integer solution.

**For** each branching constraint that was previously active in $Q^o$ and is now redundant **do**:

Suppose that the branching constraint under consideration links $Q^w$ and $Q^v$ ($Q^v$ being the parent of $Q^w$). In a *low-level simplification*, $Q^w \equiv Q^o$.

1. Cut off the branch $Q^v$–$Q^w$.
2. Link $Q^w$ directly to the parent of $Q^v$, say $Q^u$, by the branching constraint that previously linked $Q^v$ to $Q^u$; if $Q^v$ was the root then $Q^w$ becomes the root.
3. Remove $Q^v$, its upper link and its descendants from the branch opposite to $Q^w$.
4. Concerning $Q^w$ and its descendants, leave just the intermediate nodes needed to get $Q^o$ but assure that they remain forked: considering that there are $q$ intermediate nodes between $Q^w$ and $Q^o$, i.e., $Q^w \equiv Q_0, Q_1, \ldots Q_q, Q_{q+1} \equiv Q^o$, replace the descendants of $Q_i$, $i = 0,\ldots,q$, from the branch opposite to $Q_{i+1}$ with a new single node (a temporary terminal node).

Now, neither $Q^w$ nor another descendant of $Q^w$ includes in its linear programming problem the bounding constraint that linked $Q^w$ to $Q^v$.

**If** the simplified tree has consecutive branching constraints on the same variable, **then** a further simplification is made (following the steps above) in order to discard the constraint that is redundant for $Q^o$.

This additional simplification is illustrated in Fig. 6.12. Note that the new right branch of $Q^w$ in Fig. 6.12 includes the feasible sub-regions of the old right branch of $Q^u$.

Details about the implementation of the method, namely the information of the branch-and-bound tree that is preserved during the process, as well as computational results can be found in Alves and Clímaco (2000).

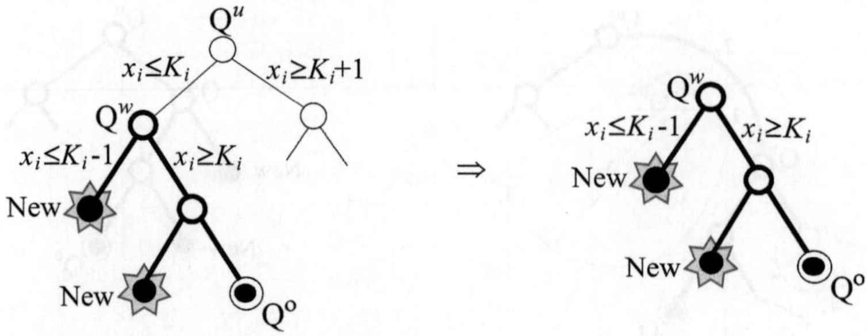


**Fig. 6.12** Additional simplification

## 6.4.2 *An Illustrative Example*

Consider the MOMILP problem (Prob. 6.1) with two integer variables, $x_1$ and $x_2$, and two continuous variables, $x_3$ and $x_4$.

$$
\begin{array}{lll}
\max & z_1 = 3x_1 + x_2 + 2x_3 + x_4 & \\
\max & z_2 = x_1 - x_2 + 2x_3 + 4x_4 & \\
\max & z_3 = -x_1 + 5x_2 + x_3 + 2x_4 & \\
\text{s.t.} & 2x_1 + x_2 + 4x_3 + 3x_4 \le 56 & \text{(Prob. 6.1)} \\
& 3x_1 + 4x_2 + x_3 + 2x_4 \le 55 & \\
& x_j \ge 0,\ j = 1,..,4 & \\
& x_1, x_2 \text{ integer} &
\end{array}
$$

The pay-off table was firstly computed. The ideal point to this problem is $\mathbf{z}^* = (60, 74.667, 68)$.

Suppose that the DM chooses the reference point $\mathbf{z}^+ = (108, 80, 75)$ to start the procedure: the scalarizing problem (6.9) (considering $\rho = 0.001$) is solved using the branch-and-bound method yielding the efficient solution $\mathbf{x} = (10, 4, 8, 0)$ whose image in the objective function space is $\mathbf{z} = (50, 22, 18)$.

Suppose that the DM wants to perform a *directional search* to improve the objective function $f_2(\mathbf{x})$. The second component of the reference point will be increased and (6.9) gives place to the parametric Chebyshev programming problem $P(\theta_2)$, i.e., (6.10) with $j = 2$.

The branch-an-bound tree that solved the scalarizing problem (6.9) is shown in Fig. 6.13. The information on each terminal node $Q^r$ relevant for the next directional search is also included (see Fig. 6.13): value of the scalarizing function ($f^r$), the dual variable of the constraint associated with the second objective function ($\pi_2^r$), the maximum value of the parameter that keeps the current basis unchanged ($\theta_j^{r,\max}$) and whether the solution satisfies all the integrality constraints (Int.) or not (NInt.).

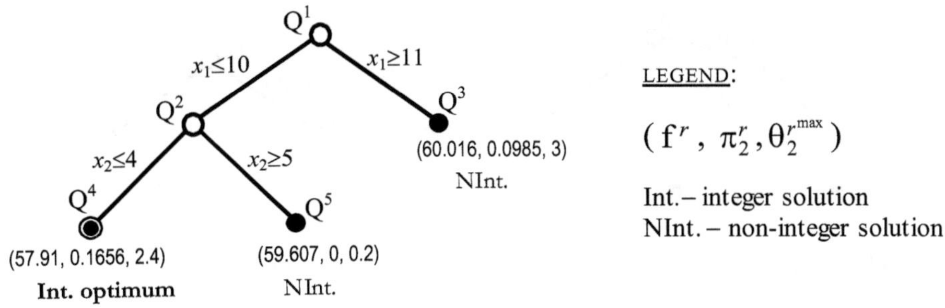


**Fig. 6.13** The Branch-and-bound tree for the reference point $\mathbf{z}^+ = (108, 80, 75)$

In this example we will consider $0 \leq \varepsilon \leq 0.1$ in $\hat{\theta}_j = \theta_j^{max} + \varepsilon$ and, for simplicity reasons, $\varepsilon$ is set such that $\hat{\theta}_j$ has only one decimal digit.

**First Iteration**

*(S) Sensitivity analysis*

$Q^4$ (in Fig. 6.13) is the current optimal node and the solution remains integer for any variation of $\theta_2$ that keeps the same basis, i.e., for $\theta_2 \leq \theta_2^{4^{max}} = 2.4$. Each terminal node $Q^r$ for which $\pi_2^r < \pi_2^4$ is compared with $Q^4$ by computing the corresponding *intersection* parameter value: $\theta_2^{4,3} = 31.386$, $\theta_2^{4,5} = 10.25$. $\theta_2^{max} = \min\{\theta_2^{4^{max}}, \theta_2^{4,3}, \theta_2^{4,5}\} = 2.4$. Hence, $Q^4$ still provides optimal solutions to the scalarizing problems for $\theta_2 \leq 2.4$. Since the current basis of $Q^4$ remains feasible for $\theta_2 \leq 2.4$ and the variation of $\theta_2$ does not destroy the integer-feasibility of the solutions, the computation of nondominated solutions closest to reference points from (108, 80, 75) to (108, 82.4, 75) is straightforward. Supposing that the DM chooses a stepsize $\mu = 1$ % (maximum normalized difference in two consecutive values of $f_2(\mathbf{x})$, the objective function being improved), a sequence of two reference points is defined by the algorithm leading to the objective function values below. As it was expected, $z_2$ is being improved with respect to the previous solution.

- $\mathbf{z}^+ = (108, 81.061, 75) \rightarrow \mathbf{z} = (49.823, 22.884, 18.442)$
- $\mathbf{z}^+ = (108, 82.122, 75) \rightarrow \mathbf{z} = (49.646, 23.768, 18.884)$.

To continue the search, the parameter value is then set slightly higher than $\theta_2^{max}$: $\hat{\theta}_2 = 2.4 + \varepsilon^{(1)} = 2.5$. The next reference point is (108, 82.5, 75).

*(U) Updating the tree for* $\mathbf{z}^+ = (108, 82.5, 75)$

Situation (U.3)—the basis of $Q^4$ changes leading to a non-integer solution. The branching constraint $x_2 \leq 4$ becomes redundant and a low-level simplification of the tree is performed (Fig. 6.14). The information on the other terminal node (only $Q^3$) is updated.

The branch-and-bound method is applied, starting with the simplified tree and continuing until the optimum of the scalarizing problem is reached (Fig. 6.15). The following nondominated solution is obtained: $\mathbf{x} = (10, 4, 7.4, 0.8)$, $\mathbf{z} = (49.6, 24, 19)$.

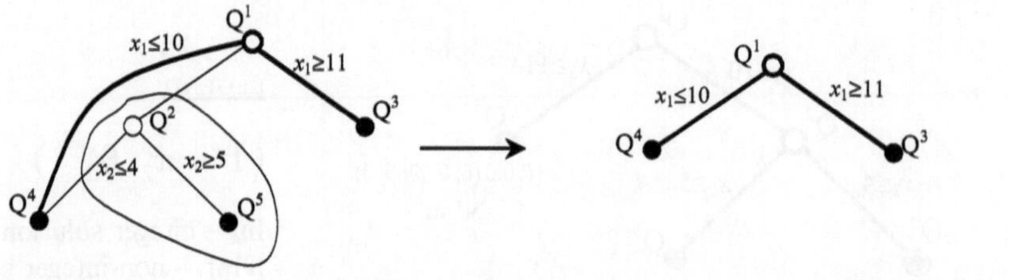


**Fig. 6.14** Simplification of the previous tree to obtain the starting-tree for $\mathbf{z}^+ = (108, 82.5, 75)$

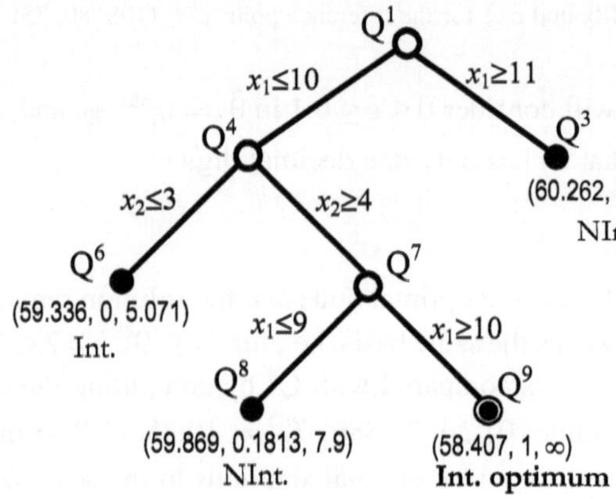


**Fig. 6.15** The final tree for $\mathbf{z}^+ = (108, 82.5, 75)$

Let us suppose that the DM wishes to continue the search along the same direction (improving $z_2$) in this and the next interactions. Hence, the procedure returns to the sensitivity analysis. Whenever the reference point is changed, $\theta_2$ is reset to 0 for the next iteration.

**Second Iteration**

*(S) Sensitivity analysis*

$Q^9$ (in Fig. 6.15) is the new current optimal node and the solution remains integer for every positive change of $\theta_2$. $\theta_2^{max} = \min\{\theta_2^{9max}, \theta_2^{9,6}, \theta_2^{9,8}, \theta_2^{9,3}\} = \min\{\infty, 0.929, *,*\} = 0.929$. The entries * do not need to be computed because $f^* > f^6$ and $\pi_2^{*} > \pi_2^{6}$ both for $Q^8$ and $Q^3$, which lead to *intersection* values larger than 0.929.

Thus, for $\theta_2 \leq 0.929$, $Q^9$ is still the optimal node of the scalarizing problem and the computation of nondominated solutions closest to reference points from (108, 82.5, 75) to (108, 83.429, 75) is straightforward because they are given by the same basis of the LP problem associated with $Q^9$.

To continue the search, $\hat{\theta}_2$ is then set to $\theta_2^{max} + \varepsilon^{(2)} = 0.929 + \varepsilon^{(2)} = 1.0$ and the new reference point is $\mathbf{z}^+ = (108, 83.5, 75)$.

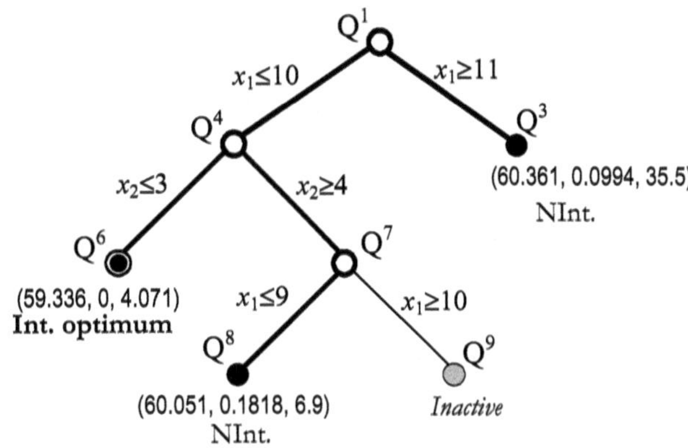


**Fig. 6.16** The final tree for $\mathbf{z}^+ = (108, 83.5, 75)$

*(U) Updating the tree for* $\mathbf{z}^+ = (108, 83.5, 75)$
Situation (U.2)—$Q^9$ is *intersected* by $Q^6$. The basis of $Q^6$ does not change for $\hat{\theta}_2$ and the corresponding solution is still integer. The tree structure is not affected; only the information on terminal nodes must be updated (Fig. 6.16). The new nondominated solution is given by $Q^6$: $\mathbf{x} = (10, 3, 6.857, 1.857)$, $\mathbf{z} = (48.571, 28.143, 15.571)$.

**Third Iteration**
*(S) Sensitivity analysis*

$Q^6$ (in Fig. 6.16) is the current optimal node with $\pi_2^6 = 0$; the $s_2$ variable (surplus variable of the constraint associated with $f_2(\mathbf{x})$ in the scalarizing problem) is basic. Hence, the previous nondominated solution remains the closest one to reference points $(108, 83.5 + \theta_2, 75)$ with $0 < \theta_2 \leq \theta_2^{6\max} = s_2 = 4.071$. So, $\hat{\theta}_2 = 4.071 + \varepsilon^{(3)} = 4.1$.

*(U) Updating the tree for* $\mathbf{z}^+ = (108, 87.6, 75)$
Situation (U.3)—the basis of $Q^6$ changes. Since $Q^6$ is the best candidate node and its new solution is integer, $Q^6$ remains the optimal node, $\mathbf{x} = (10, 3, 6.850, 1.867)$, $\mathbf{z} = (48.567, 28.167, 15.583)$.

**Next Iterations**
In order to illustrate different situations, we will skip several iterations assuming that the search has followed the same direction. Accordingly, consider the reference point $\mathbf{z}^+ = (108, 120.5, 75)$ corresponding to the tree shown in Fig. 6.17, $\mathbf{x} = (10, 0, 0.875, 10.833)$, $\mathbf{z} = (42.583, 55.083, 12.542)$. $Q^{22}$ is the current optimal node.

*(S) Sensitivity analysis*
$Q^{22}$ (in Fig. 6.17) is the current optimal node and the solution remains integer for variations of $\theta_2$ within the current basis, i.e., for $\theta_2 \leq \theta_2^{22\max} = 3.5$. In addition, no other node satisfies the requirement for checking the *intersection* condition. So, $Q^{22}$ is still the optimal node of the scalarizing problem for $\theta_2 \leq 3.5$ and nondominated solutions closest to reference points from (108, 120.5, 75) to (108,124, 75) can be easily computed by parametric linear programming applied to $Q^{22}$.

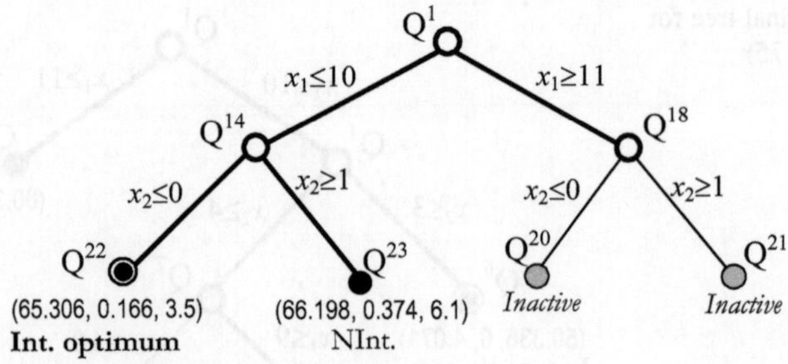


**Fig. 6.17** The final tree for $\mathbf{z}^+ = (108, 120.5, 75)$

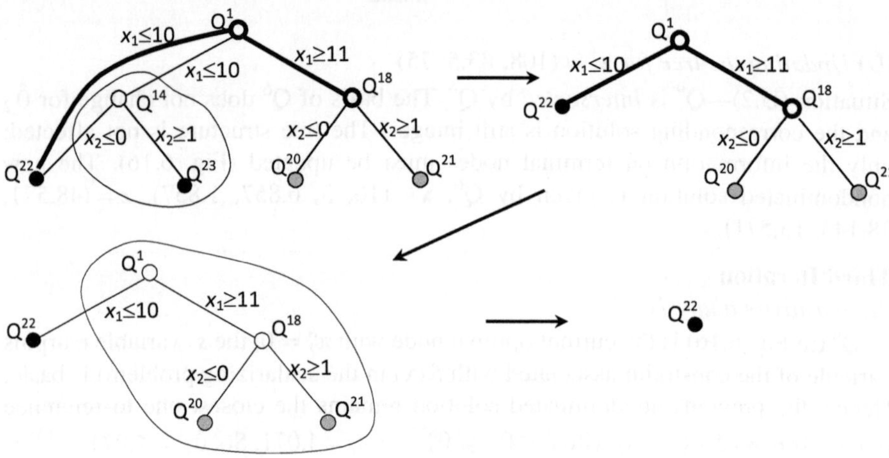


**Fig. 6.18** Simplification yielding the starting-tree for $\mathbf{z}^+ = (108, 124.1, 75)$

Let $\hat{\theta}_2 = 3.5 + \varepsilon = 3.6$ and the new reference point $\mathbf{z}^+ = (108, 124.1, 75)$.

*(U) Updating the tree for* $\mathbf{z}^+ = (108, 124.1, 75)$
Situation (U.3)—the optimal basis of $Q^{22}$ changes and the bounding constraints $x_2 \leq 0$ and $x_1 \leq 10$ become redundant for $Q^{22}$. Although the non-negativity constraint for $x_2$ is active, so that $x_2 = 0$, the solution is non-integer because $x_1 = 9.975$. $Q^{22}$ is the best candidate node and should be branched. Before branching, two simplifications of the tree are done: firstly, the bounding constraint $x_2 \leq 0$ is picked and afterwards $x_1 \leq 10$, leading to two consecutive low-level simplifications which are shown in Fig. 6.18.

Starting the branch-and-bound with the simplified tree—just the node $Q^{22}$—and continuing until the optimum of the scalarizing problem is reached (Fig. 6.19), a nondominated solution is obtained: $\mathbf{x} = (10, 0, 0, 12)$, $\mathbf{z} = (42, 58, 14)$.

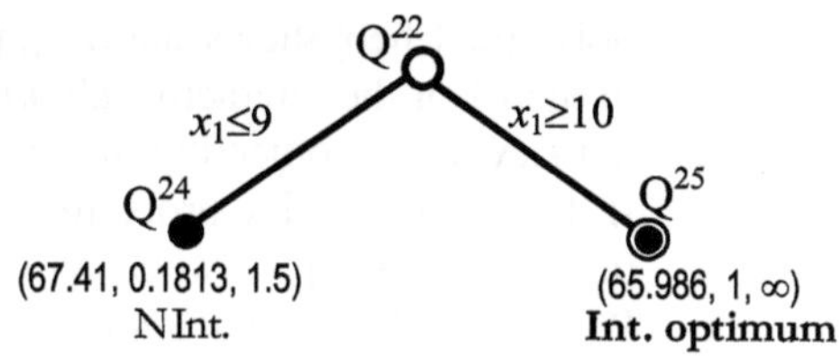


**Fig. 6.19** The final tree for $\mathbf{z}^+ = (108, 124.1, 75)$

### *6.4.3 Generating Method for Bi-Objective Problems*

The interactive reference point method presented above can be used as a generating algorithm to characterize the whole nondominated set of bi-objective MOMILP problems. It can generate all nondominated solutions to MOILP problems and, in the mixed-integer case, the user sets a stepsize μ that defines the maximum variation (in percentage) desired for the value of one of the objective functions when continuous nondominated solutions are computed. This stepsize is also used in the interactive algorithm to define the proximity of continuous nondominated solutions obtained during a directional search (see, e.g., the first iteration of the previous example).

To examine the whole nondominated set of a bi-objective problem, the algorithm can either start at the optimum of the first objective function, and perform a directional search that improves the second objective, or do the reverse. Since increasing an objective function implies decreasing the other, the nondominated set is fully determined using such an approach except for a gap between continuous solutions that is controlled by the stepsize μ. As stated above, the algorithm automatically recognizes when it reaches the nondominated solution that optimizes one objective function (even if the pay-off table has not been computed). Therefore, if a directional search is performed to improve $f_j(\mathbf{x})$ and a nondominated solution maximizing this function is at hand, then the procedure indicates that no more improvement is possible and the directional search finishes.

Without loss of generality, let us consider that the algorithm starts at the optimum of the first objective function. So, $f_1(\mathbf{x})$ is firstly maximized (using a weighted-sum of the objectives or a lexicographic optimization approach to ensure that a nondominated solution is obtained). Let this solution be $\mathbf{x}^{1*}$ with $\mathbf{z}^{1*} = (f_1(\mathbf{x}^{1*}), f_2(\mathbf{x}^{1*}))$. It can be proved that there always exists a small enough ρ such that solution $\mathbf{x}^{1*}$ also optimizes the achievement scalarizing problem (6.8) without weights (i.e., all $\lambda_k = 1$) considering $\overline{\mathbf{z}} = \mathbf{z}^{1*}$ (Alves et al. 2012). This is still true for the augmented Chebyshev programming problem (6.9). So, the first reference point to be considered is $\mathbf{z}^{1*}$.

The bi-objective generating algorithm can be stated as follows:

*Step 0.* Compute the pay-off table of the bi-objective MOMILP problem or just a nondominated solution that maximizes $f_1(\mathbf{x})$. Let $\mathbf{z}^{1*}$ be its image in the objective function space.

*Step 1.* Define the first reference point as $\mathbf{z}^+ = \mathbf{z}^{1*}$.

Solve the Chebyshev scalarizing problem (6.9) using branch-and-bound as in Step 1 of the interactive algorithm.

*Step 2.* Set $f_2(\mathbf{x})$ as the objective function to be improved.
If the multiobjective problem is mixed-integer, choose a stepsize $\mu > 0$ that defines a gap between continuous nondominated solutions.
Perform a *directional search* as in Step 3 of the interactive algorithm stopping when a nondominated solution that maximizes $f_2(\mathbf{x})$ is reached.

The stepsize μ represents the maximum value that is allowed for the ratio $\left(z_2^{new} - z_2^{prev}\right)/\left(\widetilde{z}_2^* - z_2^{1*}\right)$, where $\mathbf{z}^{new}$ and $\mathbf{z}^{prev}$ are two consecutive nondominated solutions on a continuous path, the *new* and the *previous* one, respectively, $\widetilde{z}_2^*$ is an approximation of the maximum of $f_2(\mathbf{x})$ (e.g., the maximum of $f_2(\mathbf{x})$ in the linear relaxation of the problem) or its true maximum value if the pay-off table has been computed in *Step 0*.

The algorithm above starts at the optimum of $f_1(\mathbf{x})$ and stops at the optimum of $f_2(\mathbf{x})$. Naturally, starting at the optimum of $f_2(\mathbf{x})$ and selecting the first objective to be improved is another possibility to scan the nondominated region of the bi-objective problem. In this case, μ is used for restricting differences in $f_1(\mathbf{x})$.

**Example**

Consider the following MOILP problem (Prob. 6.2) with two objective functions and two integer variables (this problem has been displayed above in Fig. 6.6).

$$\begin{aligned} \max \quad & z_1 = f_1(\mathbf{x}) = x_1 - x_2 \\ \max \quad & z_2 = f_2(\mathbf{x}) = -x_1 + 2x_2 \\ \text{s.t}: \quad & x_1 + 6x_2 \leq 21 \\ & 14x_1 + 6x_2 \leq 63 \\ & x_1,\ x_2 \geq 0 \text{ and integer} \end{aligned} \qquad \text{(Prob. 6.2)}$$

*Step 0.* The nondominated solution that maximizes $f_1(\mathbf{x})$ is computed: $\mathbf{x}^{1*} = (4,0)$, $\mathbf{z}^{1*} = (4,-4)$.

*Step 1.* The first reference point is set to $\mathbf{z}^+ = (4, -4)$. The Chebyshev scalarizing problem is solved, obtaining again $\mathbf{x}^{1*}$, $\mathbf{z}^{1*}$.

*Step 2.* The objective function $f_2(\mathbf{x})$ is selected to be improved.
(There is no need to define a stepsize μ because this is a MOILP problem).
Perform a *directional search* to improve $f_2(\mathbf{x})$:

- the reference point is changed to $\mathbf{z}^+ = (4, -3)$ yielding the solution $\mathbf{x}^2 = (4,1)$, $\mathbf{z}^2 = (3,-2)$,
- the reference point is changed to $\mathbf{z}^+ = (4, 0)$ yielding the solution $\mathbf{x}^3 = (3,1)$, $\mathbf{z}^3 = (2,-1)$,
- the reference point is changed to $\mathbf{z}^+ = (4, 2)$ yielding the solution $\mathbf{x}^4 = (3,2)$, $\mathbf{z}^4 = (1,1)$,
- the reference point is changed to $\mathbf{z}^+ = (4, 5)$ yielding the solution $\mathbf{x}^5 = (3,3)$, $\mathbf{z}^5 = (0,3)$,

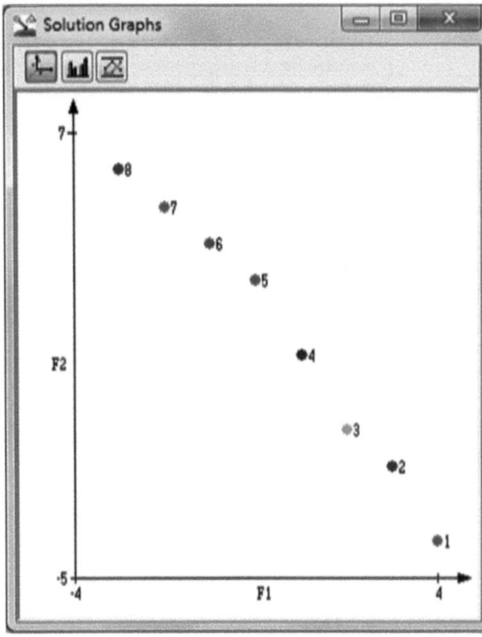


**Fig. 6.20** All nondominated solutions of (Prob. 6.2)

- the reference point is changed to $\mathbf{z}^+ = (4,\ 8)$ yielding the solution $\mathbf{x}^6 = (2,3)$, $\mathbf{z}^6 = (-1,4)$,
- the reference point is changed to $\mathbf{z}^+ = (4,\ 11)$ yielding the solution $\mathbf{x}^7 = (1,3)$, $\mathbf{z}^7 = (-2,5)$,
- the reference point is changed to $\mathbf{z}^+ = (4,\ 13)$ yielding the solution $\mathbf{x}^8 = (0,3)$, $\mathbf{z}^8 = (-3,6)$,
- indication that the optimum of $f_2(\mathbf{x})$ has been reached at solution 8.

  The eight solutions obtained constitute the set of all nondominated solutions to the problem (Prob. 6.2). These solutions are depicted in the objective function space in Fig. 6.20.

### 6.4.4 *The Software*

A software for MOMILP problems accompanies this book. It implements the interactive reference point method described in Sect. 6.4.1, also including the generating algorithm for bi-objective problems (Sect. 6.4.3) and a set of tools that can be used at any phase of the decision process. These tools aim at providing a progressive learning of solutions and a gradual establishment of the DM's preferences. Some of these tools, such as the optimization of weighted-sums of the objective functions, may be more useful in an initial phase of the decision process.

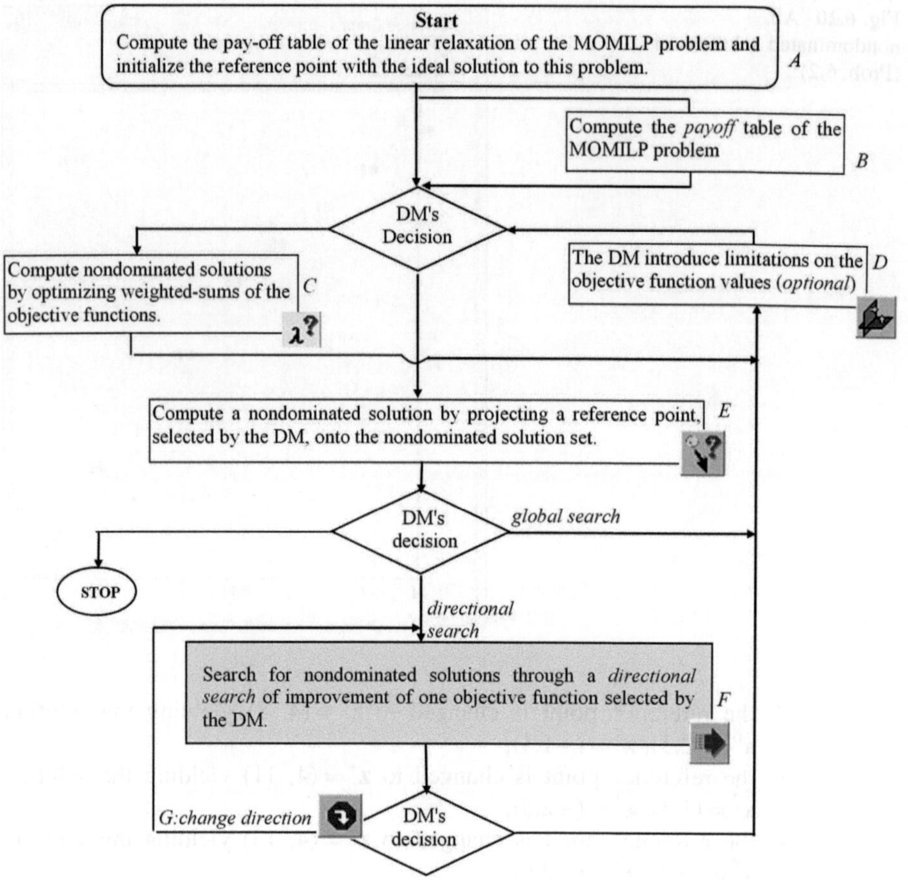


**Fig. 6.21** Scheme of interaction with the DM

The combination of the *directional search* with the possibility of imposing additional limitations on the objective function values can be used to scan nondominated solutions in different directions or to carry out a search focused on a delimited region (local search), for instance within the neighborhood of a nondominated solution that the DM considers interesting. In general, the latter option would be more useful in a final phase of the decision process.

The flowchart in Fig. 6.21 outlines the scheme of interaction with the DM implemented in this software.

The interface of the software for MOMILP is similar to the one of the *iMOLPe* software for MOLP. The main window (Fig. 6.22) includes a menu bar that provides access to any operation and icon-controls on a tool panel (with two tabs, *Standard* and *View options*) to perform specific operations, such as:

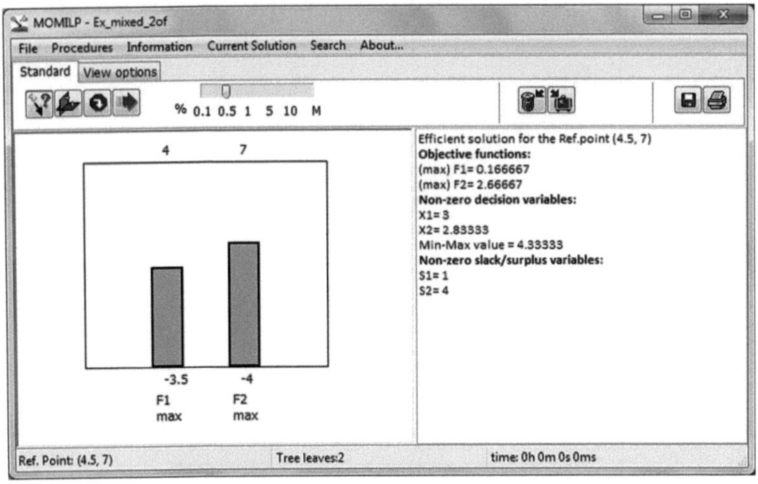


**Fig. 6.22** Main window of the software for MOMILP

- inputing preference information, namely to specify new reference points (information needed to perform *E* in Fig. 6.21), to choose an objective function to be improved during a directional search (*G*, to perform *F*), to specify weight vectors for weighted-sums (to perform *C*) or to impose bounds on the objective function values (*D*);
- requiring the computation of new nondominated solutions by projecting a reference point onto the nondominated solution set (*E*), starting a new *directional search* (entering *F*), continuing in the same directional search (keeping in *F*) or optimizing a weighted-sum (*C*); these computations can be performed considering the original feasible region or a restricted region delimited by bounds on the objective functions (*D*);
- operations concerning visualization and analysis of the results, in particular to set up display aspects (colors, number of solutions visible in the main window) or to open separate windows that show different numerical or graphical information; to insert or delete nondominated solutions into a solution archive ("bag") or to save solutions to disk.

The available controls depend on the procedure previously selected in the menu. Figure 6.22 shows the controls in the *Standard* tab when an operation *E* or *F* has been just performed. The track bar 

enables to adjust the stepsize μ and it is only available for mixed-integer problems.

Figure 6.23 shows the nondominated frontier of a mixed-integer problem, which has been scanned using the generating algorithm with a stepsize of 0.1 %. The problem is similar to (Prob. 6.2) but it has $x_1$ integer and $x_2$ continuous. Figure 6.24 shows the data of the objective functions and the constraints of this problem in the editing environment provided by the software. The type of the variables (binary, integer or continuous) and their lower and upper bounds are defined through a window as in Fig. 6.25. This figure shows the definition of $x_1$ as integer and $x_2$ as continuous, both with lower bound 0 and no upper bound (represented by M).

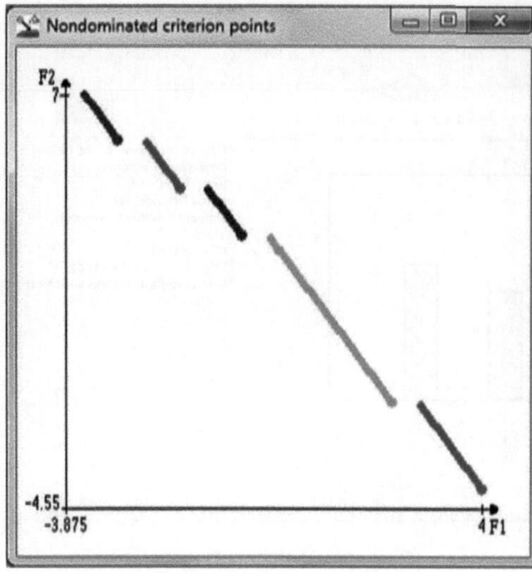


Fig. 6.23 Nondominated frontier of a mixed-integer problem

| Funct.: | X1 | X2 | max/min |
|---|---|---|---|
| [1] | 1 | -1 | max |
| [2] | -1 | 2 | max |

| Constr.: | X1 | X2 | (<,>,=) | b |
|---|---|---|---|---|
| [1] | 1 | 6 | < | 21 |
| [2] | 14 | 6 | < | 63 |

Fig. 6.24 Editing the objective functions and the constraints

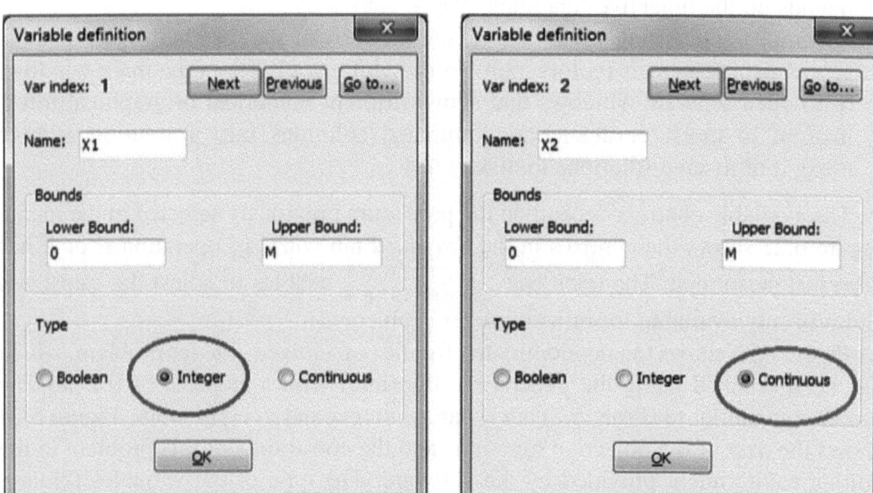


Fig. 6.25 Defining the type of the variables and their lower and upper bounds

For multiobjective all-integer problems (MOILP) with 2 or 3 objective functions, the software also includes a graphical representation of indifference regions of reference points, i.e., sets of reference points that lead to the same nondominated solution. As presented in the previous chapters, in MOLP it is possible to determine the indifference region on the weight space associated with each nondominated basic solution computed. However, indifference regions in MOILP/MOMILP are very difficult to compute. Unless all nondominated solutions are already known, there is no recognized method that enables to compute a complete indifference region at once (i.e., after only one optimization), either in the weight space (when weighted-sums of the objective functions are optimized) or in the reference point parametric space. This software calculates convex sets of reference points that lead to the same solution, which are, in general, partial rather than complete indifference regions in the reference point space. The procedure (described in Alves and Clímaco, 2001) takes advantage of the directional searches to define larger subsets of indifference regions that are successively appended. Indifference regions of reference points are usually non-convex unlike indifference regions on the weight space, which are always convex regardless the type of the problem (MOLP, MOILP, MOMILP or non-linear problem). An approach to compute subsets of indifference regions on the weight space to MOILP/MOMILP was recently proposed in (Alves and Costa 2016).

The reference point space coincides with the objective function space. Indifference regions are not lost if only a cut of this space is visualized in which the sum of the components of the reference points is constant. This means a plane for 3 objective functions and a line for 2 objective functions. Furthermore, the representation of indifference regions can be limited to the extent of a triangle/line segment provided that the limits are defined large enough so that all nondominated solutions are included in the picture; the individual optima of the objective functions fill areas near the vertices of the triangle/line segment.

Figure 6.26 shows examples of the representation of subsets of indifference regions in the reference point space for (*a*) a problem with three objective functions and (*b*) a problem with two objective functions. In the first case, the regions were obtained when the pay-off table was computed (solutions 1, 2 and 3, respectively) and a directional search was conducted from solution 4 (obtained by choosing the ideal solution as reference point) in order to improve $f_2(\mathbf{x})$; this directional search ended at solution 2, which is the optimum of $f_2(\mathbf{x})$, passing through solutions 5 and 6. The graph of Fig. 6.26b corresponds to the problem (Prob. 6.2) and was obtained at the same time as the generating algorithm was applied to the problem. In the case of bi-objective problems, the whole indifference regions are obtained when a directional search is performed.

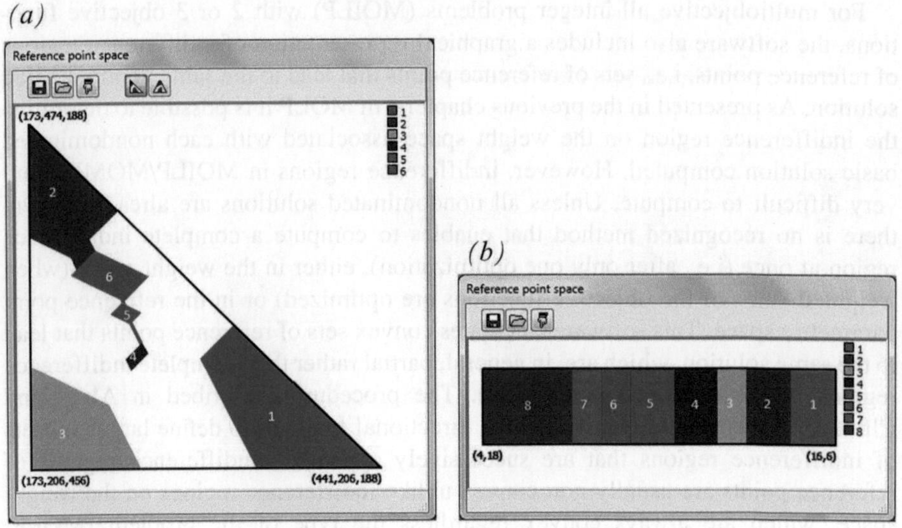


Fig. 6.26 Subsets of indifference regions in the reference point space

## 6.5 Proposed Exercises

1. Consider the following MOILP problem:

$$
\begin{aligned}
\max\ & f_1(\mathbf{x}) = x_1 \\
\max\ & f_2(\mathbf{x}) = -x_1 + 2x_2 \\
\text{s.t.}\ & \\
& x_1 \leq 5 \\
& 7x_1 + 6x_2 \leq 42 \\
& 9x_1 + 20x_2 \leq 90 \\
& x_1, x_2 \geq 0 \text{ and integer}
\end{aligned}
$$

   (a) Represent graphically the problem in the decision space and determine all efficient/nondominated solutions.
   (b) What are the nondominated solutions that are reachable using weighted-sums of the objective functions?

2. Consider the problem of exercise 1 with $x_1$ integer and $x_2$ continuous.

   (a) Identify all supported nondominated extreme and non-extreme solutions.
   (b) Identify all unsupported nondominated solutions.

3. Consider the following MOILP problem:

$$\begin{aligned}
&\max\ z_1 = f_1(\mathbf{x}) = 3x_1 + x_2 + 2x_3 + x_4\\
&\max\ z_2 = f_2(\mathbf{x}) = x_1 - x_2 + 2x_3 + 4x_4\\
&\text{s.t.}\\
&2x_1 + x_2 + 4x_3 + 3x_4 \le 56\\
&3x_1 + 4x_2 + x_3 + 2x_4 \le 55\\
&x_1, x_2, x_3, x_4 \ge 0 \text{ and integer}
\end{aligned}$$

Suppose that the following two nondominated solutions to this problem have already been computed and the DM wishes to search for other nondominated solutions between $\mathbf{z}^a$ and $\mathbf{z}^b$.

$$\begin{aligned}
\mathbf{x}^a &= (7,\ 0,\ 0,\ 14) \qquad & \mathbf{z}^a &= (35,\ 63)\\
\mathbf{x}^b &= (4,\ 0,\ 0,\ 16) \qquad & \mathbf{z}^b &= (28,\ 68)
\end{aligned}$$

(a) Formulate a weighted-sum of the objective functions with additional bounds on the objectives that enables to find a nondominated solution (if it exists) between this pair of solutions.

(b) Solve the problem formulated in a) using the MOMILP *software*.
*Hint*: impose additional constraints on the objectives by choosing the option  and then call the weighted-sum procedure. 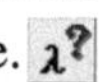


(c) Follow a similar process as the method of Chalmet et al. (Sect. 6.2) to continue the search for new nondominated solutions between the new pairs of solutions.

(d) Discuss the impact of choosing a particular weight vector when the purpose is to compute all nondominated solutions between a pair of solutions in a bi-objective integer problem.

(e) Use the MOMILP *software* to generate all nondominated solutions to this bi-objective problem.

4. Consider the problem of exercise 3 with $x_1$, $x_2$ integer and $x_3$, $x_4$ continuous.
Use the MOMILP *software*:

(a) Suppose you want to compute a nondominated solution that optimizes $f_2(\mathbf{x})$. Optimize a weighted-sum of the objectives with appropriate weights to obtain that solution.

(b) Choose the reference point procedure and an appropriate reference point to compute (again) the nondominated solution that optimizes $f_2(\mathbf{x})$. What reference point did you choose?

(c) Suppose you solved (b) without knowing neither the result of (a) nor any other nondominated solution to the multiobjective problem. Indicate a reference point that you probably would choose for (b). How could you be sure that you have found the solution that optimizes $f_2(\mathbf{x})$ using only the interactive reference point algorithm?

(d) Scan all nondominated solutions by performing a *directional search*, starting at the previous solution and choosing $f_1(\mathbf{x})$ to be improved, until the optimum of this objective function is reached.

5. Consider the following MOILP problem:

$$\begin{aligned}
\max z_1 &= f_1(\mathbf{x}) = x_1 \\
\max z_2 &= f_2(\mathbf{x}) = x_2 \\
\max z_3 &= f_3(\mathbf{x}) = x_3 \\
\text{s.t.} & \\
& 3x_1 + 2x_2 + 3x_3 \leq 18 \\
& x_1 + 2x_2 + x_3 \leq 10 \\
& 9x_1 + 20x_2 + 7x_3 \leq 96 \\
& 7x_1 + 20x_2 + 9x_3 \leq 96 \\
& x_1, x_2, x_3 \geq 0 \text{ and integer}
\end{aligned}$$

Suppose that the following two nondominated solutions to this problem are already known:

$$\begin{aligned}
\mathbf{z}^a = \mathbf{x}^a &= (4,\ 3,\ 0) \\
\mathbf{z}^b = \mathbf{x}^b &= (1,\ 0,\ 5)
\end{aligned}$$

(a) Considering the weight vector $\boldsymbol{\lambda} = (^1/_3,\ ^1/_3,\ ^1/_3)$, formulate a weighted-sum problem with additional constraints and auxiliary binary variables that ensures the computation of a nondominated solution different from $\mathbf{z}^a$ and $\mathbf{z}^b$, as in the methods of Klein & Hannan or Sylva & Crema.

(b) Can the optimal solution to the problem formulated in a) be an unsupported nondominated solution or will it certainly be a supported solution?

6. Consider the following MOILP problem with 3 integer variables and 2 binary variables:

$$\begin{aligned}
\max z_1 &= f_1(\mathbf{x}) = 8x_1 + 7x_2 + 10x_3 + 5x_4 - 2x_5 \\
\max z_2 &= f_2(\mathbf{x}) = 3x_1 - x_2 + x_3 + x_4 - x_5 \\
\max z_3 &= f_3(\mathbf{x}) = 9x_1 - x_3 + 5x_4 + 10x_5 \\
\text{s.t.} & \\
& 2x_1 + 3x_3 + 5x_4 + 2x_5 \leq 30 \\
& 5x_1 + x_2 + x_3 + 2x_5 \leq 30 \\
& 5x_1 + 2x_2 + 3x_3 + 4x_5 \leq 30 \\
& 4x_1 + 3x_2 + 5x_3 + 3x_4 + 2x_5 \leq 30 \\
& x_1, x_2, x_3 \geq 0 \text{ and integer} \\
& x_4, x_5 \in \{0, 1\}
\end{aligned}$$

Using the MOMILP *software*:

(a) Compute the pay-off table of the problem. What is the ideal solution?

(b) Determine the nondominated solution closest to the ideal solution according to the augmented Chebyshev metric (reference point procedure).

(c) Perform a *directional search* from the previous solution in order to improve $f_2(\mathbf{x})$ until its maximum is achieved.
(d) Change the direction and perform now a *directional search* to improve $f_1(\mathbf{x})$.
(e) Assume that the DM wants to impose a lower bound of 60 on $f_1(\mathbf{x})$. Thus, include the constraint $f_1(\mathbf{x}) \geq 60$. Compute the nondominated solution closest to the ideal solution considering this additional constraint. Is this solution the same as the one obtained in b)?
(f) Start from the solution obtained in e) and perform a *directional search* to improve $f_3(\mathbf{x})$ keeping the constraint $f_1(\mathbf{x}) \geq 60$. How many solutions are computed? Why?
(g) Analyze the subsets of indifference regions in the reference point space and give at least three different reference points that lead to the nondominated solution that optimizes $f_2(\mathbf{x})$ (without considering additional bounds on the objectives).

# References


Aksoy Y (1990) An interactive branch-and-bound algorithm for bicriterion nonconvex/mixed integer programming. Naval Res Logist 37:403–417

Alves MJ, Clímaco J (1999) Using cutting planes in an interactive reference point approach for multiobjective integer linear programming problems. Eur J Oper Res 117(3):565–577

Alves MJ, Clímaco J (2000) An interactive reference point approach for multiobjective mixed-integer programming using branch-and-bound. Eur J Oper Res 124(3):478–494

Alves MJ, Clímaco J (2001) Indifference sets of reference points in multi-objective integer linear programming. J Multi-Criteria Decis Anal 10(4):177–189

Alves MJ, Clímaco J (2007) A review of interactive methods for multiobjective integer and mixed-integer programming. Eur J Oper Res 180:99–115

Alves MJ, Costa JP (2009) An exact method for computing the nadir values in multiple objective linear programming. Eur J Oper Res 198:637–646

Alves MJ, Costa JP (2016) Graphical exploration of the weight space in three-objective mixed integer linear programs. Eur J Oper Res 248:72–83

Alves MJ, Dempe S, Júdice JJ (2012) Computing the Pareto frontier of a bi-objective bi-level linear problem using a multiobjective mixed-integer programming algorithm. Optimization 61 (3):335–358

Alves MJ, Antunes CH, Clímaco J (2015) Interactive MOLP explorer: a graphical-based computational tool for teaching and decision support in multi-objective linear programming models. Comput Appl Eng Educ 23(2):314–326

Antunes CH, Alves MJ, Silva AL, Clímaco JN (1992) An integrated MOLP method base package—a guided tour of TOMMIX. Comput Oper Res 19(7):609–625

Benayoun R, de Montgolfier J, Tergny J, Larichev O (1971) Linear programming with multiple objective functions: step method (STEM). Math Program 1:366–375

Bitran GR (1977) Linear multiple objective programs with zero–one variables. Math Program 13:121–139

Bitran GR (1979) Theory and algorithms for linear multiple objective programs with zero–one variables. Math Program 17(3):362–389

Bowman VJ Jr (1976) On the relationship of the Tchebycheff norm and the efficient frontier of multiple-criteria objectives. In: Thiriez H, Zionts S (eds) Multiple criteria decision making, lecture notes in economics and mathematical systems, vol 130. Springer, Berlin, pp 76–86

Chalmet LG, Lemonidis L, Elzinga DJ (1986) An algorithm for the bi-criterion integer programming problem. Eur J Oper Res 25:292–300




C. Henggeler Antunes et al., *Multiobjective Linear and Integer Programming*,
EURO Advanced Tutorials on Operational Research,
DOI 10.1007/978-3-319-28746-1

Chankong V, Haimes Y (1983) Multiobjective decision making: theory and methodology. North-Holland, New York

Clímaco, J, Almeida, AT (1981) Multiobjective power systems generation planning. In: Proceeding of the 3rd international conference energy use management, Pergamon Press, Oxford

Clímaco J, Antunes CH (1987) TRIMAP—an interactive tricriteria linear programming package. Found Control Eng 12:101–119

Clímaco J, Antunes CH (1989) Implementation of an user friendly software package—a guided tour of TRIMAP. Math Comput Model 12:1299–1309

Clímaco J, Ferreira C, Captivo ME (1997) Multicriteria integer programming: an overview of the different algorithmic approaches. In: Clímaco J (ed) Multicriteria analysis. Springer, Berlin, pp 248–258

Coello CAC, Van Veldhuizen DA, Lamont GB (2002) Evolutionary algorithms for solving multi-objective problems, vol 242. Kluwer Academic, New York

Cohon J (1978) Multiobjective programming and planning. Academic, New York, NY

Cohon JL, Church RL, Sheer DP (1979) Generating multiobjective tradeoffs: an algorithm for bicriterion problems. Water Resour Res 15:1001–1010

de Samblanckx S, Depraetere P, Muller H (1982) Critical considerations concerning the multicriteria analysis by the method of Zionts and Wallenius. Eur J Oper Res 10:70–76

Deb K (2001) Multi-objective optimization using evolutionary algorithms, vol 16. Wiley, Chichester

Deckro RF, Winkofsky EP (1983) Solving zero–one multiple objective programs through implicit enumeration. Eur J Oper Res 12:362–374

Durso A (1992) An interactive combined branch-and-bound/Tchebycheff algorithm for multiple criteria optimization. In: Goicoechea A, Duckstein L, Zionts S (eds) Multiple criteria decision making: theory and applications in business, industry and government. Springer, New York, pp 107–122

Evans J, Steuer R (1973) A revised simplex method for multiple objective programs. Math Program 5(1):54–72

Ferreira C, Clímaco J, Paixão J (1994) The location covering problem: a bicriterion interactive approach. Investigación Operativa 4(2):119–139

Feyerabend P (1975) Against method. New Left Books, London

French S (1984) Interactive multi-objective programming: its aims, applications and demands. J Oper Res Soc 35:827–834

Geoffrion AM (1967) Solving bi-criterion mathematical programs. Oper Res 15(1):39–54

Geoffrion AM (1968) Proper efficiency and the theory of vector maximization. J Math Anal Appl 22:618–630

Geoffrion A (1971) Duality in nonlinear programming: a simplified applications-oriented development. SIAM Rev 13(1):1–37

Geoffrion A (1983) Can Management Science/Operations Research evolve fast enough? Interfaces 13(1):10–25

Gonzalez JJ, Reeves GR, Franz LS (1985) An interactive procedure for solving multiple objective integer linear programming problems. In: Haimes YY, Chankong V (eds) Decision making with multiple objectives, lecture notes in economics and mathematical systems, vol 242. Springer, Berlin, pp 250–260

Greco S, Ehrgott M, Figueira JR (eds) (2005) Multiple criteria decision analysis: state of the art surveys. Springer, New York, NY

Greco S, Ehrgott M, Figueira JR (eds) (2016) Multiple criteria decision: state of the art surveys, 2nd edn. Springer, New York, NY

Hitch C (1953) Suboptimization in operations problems. J Oper Res Soc Am 1(3):87–99

Hwang C, Masud A (1979) Multiple objective decision making—methods and applications, vol 164, Lecture notes in economics and mathematical systems. Springer, Berlin, Heidelberg

Karaivanova J, Korhonen P, Narula S, Wallenius J, Vassilev V (1995) A reference direction approach to multiple objective integer linear programming. Eur J Oper Res 81:176–187

Karwan MH, Zionts S, Villarreal B, Ramesh R (1985) An improved interactive multicriteria integer programming algorithm. In: Haimes YY, Chankong V (eds) Decision making with multiple objectives, lecture notes in economics and mathematical systems, vol 242. Springer, Berlin, pp 261–271

Keen P (1977) The evolving concept of optimality. In: Starr M, Zeleny M (eds) TIMS studies in management science, vol 6. North Holland Publishing, Amsterdam, pp 31–57

Kirlik G, Sayın S (2014) A new algorithm for generating all nondominated solutions of multiobjective discrete optimization problems. Eur J Oper Res 232(3):479–488

Kiziltan G, Yucaoglu E (1983) An algorithm for multiobjective zero–one linear programming. Manag Sci 29(12):1444–1453

Klein D, Hannan E (1982) An algorithm for the multiple objective integer linear programming problem. Eur J Oper Res 9:378–385

Koopmans TC (1951) Analysis of production as an efficient combination of activities. In: Koopmans TC (ed) Activity analysis of production and allocation, vol 13. Wiley, New York, pp 33–37

Korhonen P (1987) VIG—a visual interactive support system for multiple criteria decision making. JORBEL 27(1):3–15

Korhonen P, Laakso J (1986a) A visual interactive method for solving the multiple criteria problem. Eur J Oper Res 24(2):277–287

Korhonen P, Laakso J (1986b) Solving generalized goal programming problems using a visual interactive approach. Eur J Oper Res 26(3):355–363

Korhonen P, Wallenius J (1988) A Pareto race. Naval Res Logist 35:615–623

Kuhn H, Tucker A (1951) Nonlinear programming. In: Neyman J (ed) Proceedings of the 2nd Berkley symposium on mathematics statistics and probability. University of California Press, Berkeley, CA, pp 481–492

Lewandowski A, Wierzbicki A (1988) Aspiration based decision analysis and support. Part I: theoretical and methodological backgrounds. Working paper WP-88-03, IIASA, Laxenburg, Austria

L'Hoir H, Teghem J (1995) Portfolio selection by MOLP using an interactive branch and bound. Found Comput Decis Sci 20(3):175–185

Lin JG (1976) Maximal vectors and multi-objective optimization. J Optim Theory Appl 18:41–64

Lokman B, Köksalan M (2013) Finding all nondominated points of multi-objective integer programs. J Glob Optim 57(2):347–365

Marcotte O, Soland RM (1980) Branch and bound algorithm for multiple criteria optimization. University of Cornell and George Washington, Ithaca, NY

Marcotte O, Soland RM (1986) An interactive branch-and-bound algorithm for multiple criteria optimization. Manag Sci 32(1):61–75

Mavrotas G, Diakoulaki D (1998) A branch and bound algorithm for mixed zero–one multiple objective linear programming. Eur J Oper Res 107:530–541

Mavrotas G, Diakoulaki D (2005) Multi-criteria branch and bound: a vector maximization algorithm for mixed 0–1 multiple objective linear programming. Appl Math Comput 171 (1):53–71

Narula SC, Vassilev V (1994) An interactive algorithm for solving multiple objective integer linear programming problems. Eur J Oper Res 79:443–450

Özlen M, Azizoğlu M (2009) Multi-objective integer programming: a general approach for generating all non-dominated solutions. Eur J Oper Res 199:25–35

Özpeynirci O, Köksalan M (2010) An exact algorithm for finding extreme supported nondominated points of multiobjective mixed integer problems. Manag Sci 56(12):2302–2315

Payne H, Polak E, Collins D, Meisel W (1975) An algorithm for bicriterion optimization based on the sensitivity function. IEEE Trans Autom Control AC-20:546–548

Przybylski A, Gandibleux X, Ehrgott M (2010) A recursive algorithm for finding all nondominated extreme points in the outcome set of a multiobjective integer programme. INFORMS J Comput 22(3):371–386

Ramesh R, Zionts S, Karwan MH (1986) A class of practical interactive branch and bound algorithms for multicriteria integer programming. Eur J Oper Res 26:161–172

Ramesh R, Karwan M, Zionts S (1989) Interactive multicriteria linear programming: an extension of the method of Zionts and Wallenius. Naval Res Logist 36(3):321–335

Ramesh R, Karwan MH, Zionts S (1990) An interactive method for bicriteria integer programming. IEEE Trans Syst Man Cybern 20(2):395–403

Rasmussen LM (1986) Zero–one programming with multiple criteria. Eur J Oper Res 26:83–95

Reeves G, Franz L (1985) A simplified interactive multiple objective linear programming procedure. Comput Oper Res 12(6):589–601

Romero C (1991) Handbook of critical issues in goal programming. Pergamon, New York, NY

Roy B (1987) Meaning and validity of interactive procedures as tools for decision making. Eur J Oper Res 31:297–303

Shin WS, Allen DB (1994) An interactive paired comparison method for bicriterion integer programming. Naval Res Logist 41:423–434

Soland RM (1979) Multicriteria optimization: a general characterization of efficient solutions. Decision Sci 10:26–38

Solanki R (1991) Generating the noninferior set in mixed integer biobjective linear programs: an application to a location problem. Comput Oper Res 18(1):1–15

Steuer R (1977) An interactive multiple objective linear programming procedure. TIMS Stud Manag Sci 6:225–239

Steuer R (1986) Multiple criteria optimization: theory computation and application. Wiley, New York, NY

Steuer RE, Choo E-U (1983) An interactive weighted Tchebycheff procedure for multiple objective programming. Math Program 26:326–344

Sylva J, Crema A (2004) A method for finding the set of non-dominated vectors for multiple objective integer linear programs. Eur J Oper Res 158:46–55

Teghem J, Kunsch PL (1986) Interactive methods for multi-objective integer linear programming. In: Fandel G, Grauer M, Kurzhanski A, Wierzbicki AP (eds) Large-scale modelling and interactive decision analysis, lecture notes in economics and mathematical systems, vol 273. Springer, Berlin, pp 75–87

Vanderpooten D (1989) The interactive approach in MCDA: a technical framework and some basic conceptions. Math Comput Model 12:1213–1220

Vanderpooten D, Vincke P (1989) Description and analysis of some representative interactive multicriteria procedures. Math Comput Model 12:1221–1238

Vassilev V, Narula SC (1993) A reference direction algorithm for solving multiple objective integer linear programming problems. J Oper Res Soc 44(12):1201–1209

Villarreal B, Karwan MH, Zionts S (1980) An interactive branch and bound procedure for multicriterion integer linear programming. In: Fandel G, Gal T (eds) Multiple criteria decision making theory and application, lecture notes in economics and mathematical systems, vol 177. Springer, Berlin, pp 448–467

Vincent T, Seipp F, Ruzika S, Przybylski A, Gandibleux X (2013) Multiple objective branch and bound for mixed 0–1 linear programming: corrections and improvements for the biobjective case. Comput Oper Res 40(1):498–509

Vincke P (1992) Multicriteria decision aid. Wiley, New York, NY

Von Neumann J, Morgenstern O (1947) Theory of games and economic behavior. Princeton University Press, Princeton, NJ

White DJ (1985) A multiple objective interactive Lagrangean relaxation approach. Eur J Oper Res 19:82–90

Yu P-L (1974) Cone convexity, cone extreme points and nondominated solutions in decision problems with multiobjectives. J Optim Theory Appl 14(3):319–376

Yu P-L, Zeleny M (1975) The set of all nondominated solutions in linear cases and a multicriteria simplex method. J Math Anal Appl 49(2):430–468

Zeleny M (1974) Linear multiobjective programming. Springer, New York, NY

Zeleny M (1982) Multiple criteria decision making. McGraw-Hill, New York, NY
Zionts S, Wallenius J (1976) An interactive programming method for solving the multiple criteria problem. Manag Sci 22:652–663
Zionts S, Wallenius J (1980) Identifying efficient vectors: some theory and computational results. Oper Res 28:785–793
Zionts S, Wallenius J (1983) An interactive multiple objective linear programming method for a class of underlying nonlinear utility functions. Manag Sci 29(5):519–529

MIX
Papier aus verantwortungsvollen Quellen
Paper from responsible sources
FSC® C105338

If you have any concerns about our products,
you can contact us on
**ProductSafety@springernature.com**

In case Publisher is established outside the EU,
the EU authorized representative is:
**Springer Nature Customer Service Center GmbH**
**Europaplatz 3, 69115 Heidelberg, Germany**

Printed by Libri Plureos GmbH
in Hamburg, Germany